China's Sustainability Transitions

Ali Cheshmehzangi · Hengcai Chen

China's Sustainability Transitions

Low Carbon and Climate-Resilient Plan for Carbon Neutral 2060

Springer

Ali Cheshmehzangi
University of Nottingham Ningbo China
Ningbo, China

NERPS
Hiroshima University, Hiroshima, Japan

Hengcai Chen
University of Nottingham Ningbo China
Ningbo, China

ISBN 978-981-16-2620-3 ISBN 978-981-16-2621-0 (eBook)
https://doi.org/10.1007/978-981-16-2621-0

This Springer imprint is published by the registered company Springer Nature Singapore Pte Ltd.
The registered company address is: 152 Beach Road, #21-01/04 Gateway East, Singapore 189721, Singapore

Ali Cheshmehzangi dedicates this book to all his respected and noble teachers. They are the true inspiration for us to become educators.

Hengcai Chen dedicates this book to his parents, Mr. Chen and Mrs. Wang, and sincerely appreciates their selfless support.

The authors collectively dedicate this book to the future of cities and nations in achieving or progressing towards sustainable development. We wish them 'triumph' in their achievements, 'transparency' in their paths, and genuine 'transitions' that could make positive impacts on future generations.

Preface

This book is the first monograph on the topic of China's carbon neutrality plan. Here, we explore barriers, challenges, opportunities, and progresses that are part of China's ongoing low carbon development, future net-zero pathways, and ultimate carbon neutrality goals. Like other countries, China's next steps are challenging. Therefore, we set out some critical views, positive suggestions, and potential directions to fast-forward the next steps of carbon reductions, GHG emission reductions, and innovative solutions for a carbon neutral China. We find today's challenges to become larger challenges of tomorrow. This is if they are not resolved, or they become neglected through the never-ending cycle of growth and development.

This book highlights some key lessons from other countries for China and from China for other countries. In particular, we highlight the impacts of socio-economic and environmental crises on humanity and the planet. In doing so, we explore sustainability transitions that are low carbon and climate-resilient against the current crises and prolonged climate change challenges. With robust willingness and innovation, we believe new pathways could develop to become success stories of the next few decades. We are part of a history that could become humanity's long-term pride and a generous gift that should be passed to the next generations—that is, to achieve the true ideals of sustainable development.

China's Sustainability Transitions is vital to its future economic and social development and environmental and climate change strategies. In recent years, China's role at the global level has become more significant, and we expect it to be even more so in the coming years. For instance, the Belt and Road Initiative (BRI) is now identified as the largest infrastructure development in human history. Therefore, we can argue that China, being the main actor and initiator of this global initiative, plays a major part in the future stages of global development. It has already become a role model for many countries of the global south and has become more influential than ever before. China's recent attempts to low carbon and climate-resilient development are critical for the next steps of sustainability transitions. These would continue to happen in structured processes and systematic methods for enabling carbon reductions and towards sustainable and high-quality urbanisation. The next steps of the deep decarbonisation plan and towards carbon neutrality will be challenging but vital to China and the globe. In this regard, we value the importance of comprehensive thinking in the forthcoming steps.

Since the decarbonisation plan has become a major global agenda, cities and communities have become the main focus areas. Broadly speaking, we have no better options than shifting our current modes of development towards low carbon and then carbon neutral futures. China's paradigm shift, pathways towards changes in energy systems, and its many transitions could become a unique model of sustainable development. Such transition's speed and pace are no longer a country-level agenda but indeed a global climate action plan, which requires more attention by multiple stakeholders, all the way from the political bodies to the general public.

Undoubtedly, the next steps are going to be very challenging. However, if we set today's path in the right direction, we could realise the values of sustainability transitions and those that could lead towards a carbon neutral society. While we are far from such achievements, we believe we already have the capacity and capability of achieving a carbon neutral globe. To discuss these further, we delve into generalities and details of low carbon and climate-resilient pathways and those that could lead us towards the ultimate goal of carbon neutrality. If we fail in the next steps, we may struggle to overcome the mounting challenges and fix today's prolonged problems. If we succeed, we might re-write humanity with better prosperity, enhanced health, and sustainable futures.

Ningbo, China/Hiroshima, Japan Ali Cheshmehzangi
Ningbo, China Hengcai Chen

Acknowledgements

We would like to thank our project interns for the project titled '*Low carbon and Climate-resilient Progress in China (LCPC)*', including Mr. Haowei Lan, Ms. Yuyuan Chen, and Mr. Yida Chen. All three are from the Department of Architecture and Built Environment at the University of Nottingham Ningbo China. We also show our special appreciation to Ms. Li Xing from Xiamen University, China, for her support in data collection and data analysis.

For Chaps. 9–11, Ali Cheshmehzangi would like to thank the Asian Development Bank (ADB) and China's state agency 'National Development Reform Commission (NDRC)' for their support on an earlier project on low carbon and climate-resilient urban development in China (2017–18). In particular, he sends his sincere gratitude to Mr. Anders Pettersson from ADB in Manila, Dr. Hailong Li from the Chinese Society for Urban Studies (CSUS) in Beijing, and Dr. Xiu Yang from the National Center for Climate Change Strategy and International Cooperation (NCSC) in Beijing. We also send our thanks to all interns and research assistants who participated in our successful project, which was completed in 2018. The research assistants were from three project partners at UNNC, CSUS, and NSCS. We are grateful that our findings and recommendations were included in China's 14th Five-Year Plan.

Many other colleagues and friends have provided valuable support and input on various documents, policy reports, and governmental data. We are very grateful to them. We hope to continue working with like-minded colleagues who believe in the value of sustainability in the future steps of development and progression. We also thank Ayra and others in the same generation, who we believe are what the future is meant to be, i.e. hopefully, a more sustainable and low carbon future, where health and happiness are the backbones of our success. We continue to observe how this may turn out to be.

Ali Cheshmehzangi acknowledges the National Natural Science Foundation of China (NSFC) for funding project numbers 71850410544 and 71950410760, used for material purchase and recruitment of local research interns in China. He also sends his appreciation to the Network for Education and Research on Peace and Sustainability (NERPS) at Hiroshima University, Japan, and the Ministry of Education,

Culture, Sports, Science, and Technology (MEXT), Japan, for the recent opportunity in studying peace and sustainability. In this recently funded programme, Ali and his team study ICT-mediated platforms for smart-resilient cities and communities. Part of the work is focused on urban resilience, and climate-resilient studies cannot be neglected. Therefore, this book is regarded as one of the early outcomes of this funded project.

Contents

About the Authors

Ali Cheshmehzangi is a Professor of Architecture and Urban Design with a Ph.D. Degree in Architecture and Urban Design, a Master's Degree (M.Arch.) in Urban Design, a Graduate Certificate in Professional Studies in Architecture, and a Bachelor Degree (B.A. Hons.) in Architecture. He is an urbanist and urban designer by profession and by heart. He studies cities and city transitions, sustainable urbanism, and integrated urban design strategies. Ali is Head of the Department of Architecture and Built Environment, Director of the Centre for Sustainable Energy Technologies (CSET), and Director of Urban Innovation Lab at the University of Nottingham Ningbo China (UNNC). He is also a Specially-Appointed Professor at Network for Education and Research on Peace and Sustainability (NERPS), Hiroshima University, Japan. Currently, he works on two research projects on 'Integrated Urban Modelling Framework', and 'ICT-based smart technologies for resilient cities'. Some of his previous projects are: 'smart eco-cities in China and Europe', 'low carbon town planning in China', 'green infrastructure of cities', 'nature-based solutions in China', 'toolkit for resilient cities', 'sponge city program' and 'green development in China', 'low carbon and climate-resilient planning', and other urban transition studies. So far, Ali has +105 published journal papers and seven other published books. His books are titled '*Designing Cooler Cities*' (2017), the award-winning '*Urban Memory in City Transitions*Eco-development in China' (2018), '*Sustainable Urban Development in the Age of Climate Change*' (2019), '*Identity of Cities and City of Identities*' (2020), '*The City in Need*' (2020), '*Urban Memory in City Transitions*' (2021), and '*Sustainable Urbanism in China*' (2021). He is currently editing a volume on '*Green Infrastructure in Chinese Cities*'.

Hengcai Chen is a Research Associate in Urban Studies at the Department of Architecture and Built Environment, University of Nottingham Ningbo China (UNNC). He obtained a Master of Engineering (M.Eng.) in Urban and Regional Planning and a Bachelor of Science (B.Sc.) in Urban and Rural Planning from Xiamen University and Fuzhou University, Fujian, China, respectively. He is an enthusiastic and passionate urban planner who is keen to explore new ideas in urban-related research. So far, his research interest is centred on overarching areas of urban sustainability and low carbon city. Hengcai is currently engaged in several urban-related research studies while preparing to continue his further education at the University of Nottingham Ningbo China. His goal is to continue research on urban sustainability studies, especially in the context of China.

Chapter 1
Sustainable Development and Climate Change

1.1 Overview of Global Movements: Urgent Agenda

This book comes soon after China pledges to become a carbon neutral country by 2060. The focus here is on 'sustainability transitions', as we explore global movements and China's position in this progress. What we see in the sustainable development stream of research is enthralling (Turner, 1995; Munasinghe, 2001; Soyez & Graßl, 2008; Banuri, 2009; Chatterjee, 2011; Lawn, 2016; Butters et al., 2020). However, our progress is still limited, and we have much to do to reach the ultimate sustainability goals, such as those embedded in the Sustainable Development Goals (SDGs) (Cheshmehzangi & Dawodu, 2019). In the urbanising era that we are in, China's position has become ever important. By studying its progress and future directions, we could look into larger-scale opportunities, influences, and replicated modes of development across the globe. In doing so, we are optimistic and critical, but we aim to be progressive and enlightening, rather than cynical and deconstructive. As we are entering the challenging decade of the 2020s, under the shadow of the ongoing dreadful pandemic, we have many sustainable goals that are yet to be achieved. Several years ago, we were more optimistic about reaching the targets set by the United Nations. Still, now we are expecting to see larger-scale socio-economic downfalls and environmental crises. These may put us on hold for a while, but we ought to continue developing pathways towards sustainable development.

1.2 Key Climate Change Issues and Global Impacts

There is plenty of compelling evidence that human activities are exacerbating changes in the global climate. There has not been such rapid climate change in the last few hundred years or even over the previous thousand years. The increase of gases such as carbon dioxide leads to the earth's continuous warming, and the global average temperature and sea-level rise faster and faster. The rate of global warming since the

A. Cheshmehzangi and H. Chen, *China's Sustainability Transitions*,
https://doi.org/10.1007/978-981-16-2621-0_1

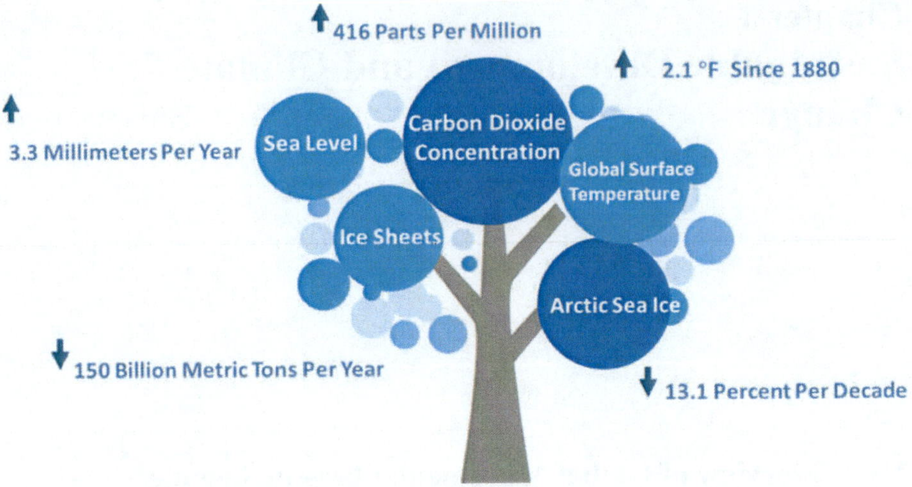

Fig. 1.1 Key climate change signs (drawn by the Authors, Data is extracted from https://climate.nasa.gov)

mid-twentieth century is unprecedented in history, and scientific evidence suggests that it is more than 95% likely to be the result of human activities (NASA, 2021a). The United States National Aeronautics and Space Administration (NASA) focuses on five key indicators or signs to describe global climate change, namely: (1) Carbon Dioxide Concentration, (2) Global Surface Temperature, (3) Arctic Sea Ice, (4) Ice Sheets, and (5) Sea Level (ibid) (Fig. 1.1).

The planet's average surface temperature had risen about 2.12° Fahrenheit (1.18 °C) since the late nineteenth century (or the year 1880 when they started recording such measures) (NASA, 2021a). The impacts of this temperature increase are severe, and we could say we are now in the era of climate change impacts. For instance, the temperature increase of just under 130 years results in the significant mass deduction of the Greenland and Antarctic ice sheets and the spring snow covered in the Northern Hemisphere. Other severe examples are global sea levels that have risen by nearly 20 cm (about 8 inches, or 0.2 m, nearly 178 mm) (NASA, 2021a) in the last century while the number of record high-temperature events has increased.

The global warming trend observed since the mid-twentieth century was attributed to the anthropogenic expansion of the greenhouse effect. For instance, as one of the greenhouse gases, Carbon dioxide (CO_2) is an important heat-trapping gas. Over the past 170 years, humans have increased atmospheric CO_2 concentration by 47% since the Industrial Revolution began. Human-produced greenhouse gases such as carbon dioxide, methane, and nitrous oxide have caused much of the observed increase in Earth's temperatures. By calculating the monthly and annual relative temperature on the base year (Moving Averages Centred/ Basic data: Jan. 1951–Dec. 1980 absolute temperature: 8.60 + /– 0.05). It can be seen that the temperature has fluctuated but continuously rose since 1994 (Fig. 1.2). All evidence responds to changes in greenhouse gas levels caused by human activities such as deforestation, land-use

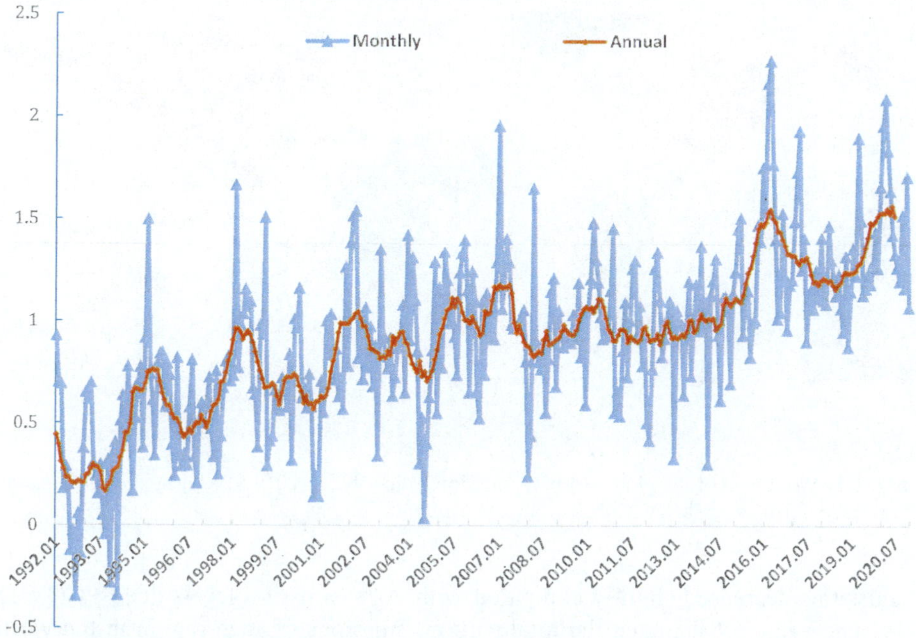

Fig. 1.2 Monthly and annual relative temperatures on the base year (drawn by the authors, data is extracted from monthly temperature records: http://berkeleyearth.org/)

changes, and burning fossil fuels. The Intergovernmental Panel on Climate Change (IPCC) forecasts a temperature rise of 2.5–10° Fahrenheit over the next century (IPCC, 2014).

The global impact of climate change is enormous and maybe result in irreversible consequences. Since 1977, the global temperature anomaly (TA) has been positive and increasing every year (Cheshmehzangi, 2020a). According to NASA reports, 2020 was the warmest year on record (NASA, 2021b), and the last decade was the warmest decade of all recorded times. These records confirm the ongoing trend of the global warming crisis and its larger-scale implications on our habitats over the last two decades. The global sea level is projected to rise another 1 to 8 feet (equivalent to 0.3 to 2.4 m) by the year 2100 and will not stop in 2200 (USGCRP, 2017). Temperatures are projected to continue rising with a reduction of soil moisture, the increasing length of the frost-free season, which exacerbates heat waves and more Droughts (Cheshmehzangi, 2020a). Therefore, ocean waters will continue to warm, and sea levels will continue to rise for many centuries at rates equal to or higher than those of the current century. The average global land and ocean surface temperature for November 2020 was 1.75° Fahrenheit, which is the second-highest November temperature on record (exceeds the now third-highest average temperature observed for November 2019) (NOAA, 2020). Global GHG emissions continued to grow and reached a record high of 52.4 GtCO2e (range: ± 5.2) without land-use change (LUC) emissions and 59.1 GtCO2e (range: ± 5.9) in 2019. In 2020, CO_2

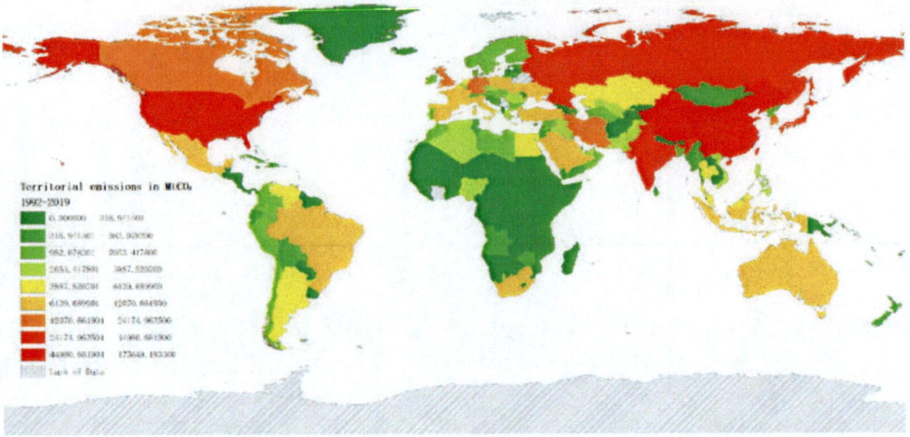

Fig. 1.3 Word map of the global carbon emission from 1992 to 2019 (drawn by the authors, data is extracted from http://www.globalcarbonatlas.org/en/CO2-emissions)

emissions decreased slightly compared with 2019 emission levels due to COVID-19. The authors calculated the total carbon emissions of each region in the world from 1992 to 2019, and obtained the results in Fig. 1.3. Over the last decade, the top four emitters (China, the United States of America, EU27 + UK, and India) have contributed to 55 percent of the total GHG emissions (UNEP, 2020).

By comparing the trends of carbon dioxide emissions and global temperature growth between 1992 and 2020, it can be seen that the increase in global temperature caused by massive emissions is quite significant (Fig. 1.4). Global emissions of greenhouse gases continue to grow. They have increased by 50% since 1990 and by 35% since 2000, driven by economic growth and fossil energy use (OECD, 2020). Climate change caused by economic and population growth affects human and natural systems and leads to many problems about economic development, natural resources, and poverty alleviation. How to deal with climate change has become an urgent issue to achieve sustainable development.

As the first published English book on the new topic of China's carbon neutrality plan, we have to cover many areas of the recent past, the present, and the forthcoming future. This chapter serves as an introduction to the book's following chapters, highlighting mainly the global progress on sustainable and climate-resilient development (Cheshmehzangi et al., 2018). The focus here is on low carbon development, decarbonisation processes, and sustainability transitions. In the following sections, we delve into general climate change challenges, starting from climate-resilience complexities and then foreseeing global challenges. Later, we highlight the importance of cities in facing the climate-resilience challenges before exploring decarbonisation progress and movements. In doing so, we look at statements by leading organisations, especially at the global level, and we briefly discuss urban responses to climate change crises. Afterward, we go into details of carbon reduction co-benefits,

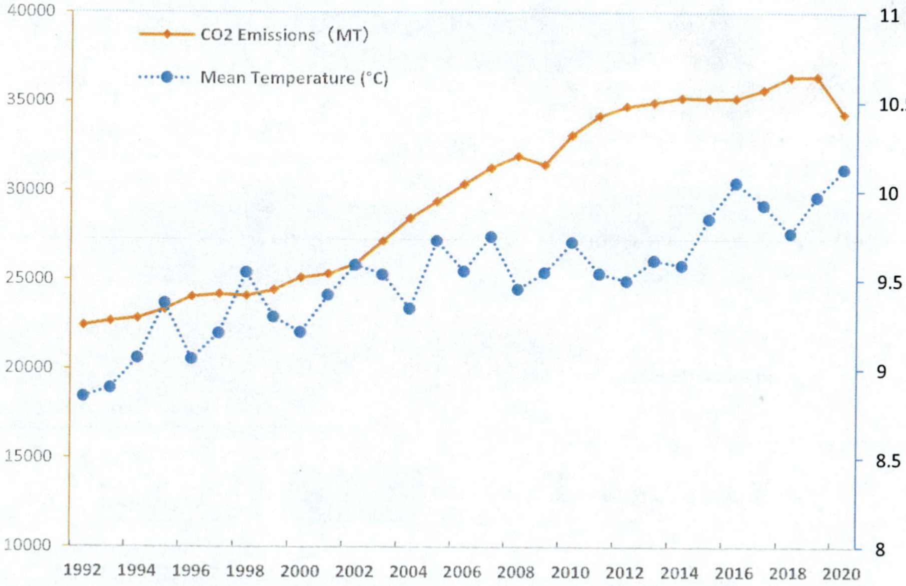

Fig. 1.4 Global temperature changes and carbon dioxide emissions (drawn by the authors, data extracted from https://data.worldbank.org)

exploring their diversity and significance regarding overacting climate change issues and solutions of cleaner productions and a healthier planet.

1.3 Climate Change and Cities: Complex Climate-Resilience Challenges

Cities are the primary consumption of energy resource, and at the same time, the focus of energy-saving and emission reduction. Over the past decade, efforts to mitigate climate change at the urban level have become more prominent. The proportion of greenhouse gas emissions from urban activities has been increasing due to the trend of global urbanisation. As the research on urban responses to climate change has grown, the International Energy Agency suggests that cities may be responsible for up to 75% of global carbon dioxide emissions from anthropogenic sources. Studies on the urban heat island effect have found that the heat released by human activities on urban land can also cause a significant increase temperature at night. (Mccarthy et al., 2010).

Cities can provide solutions through economies of scale and increased efficiency. On the other hand, unreasonable urban planning and design (for example, green infrastructure to reduce the density of the extreme, a waste of resources, gases, and pollutants release) can not only increase emissions. They also aggravate the threats

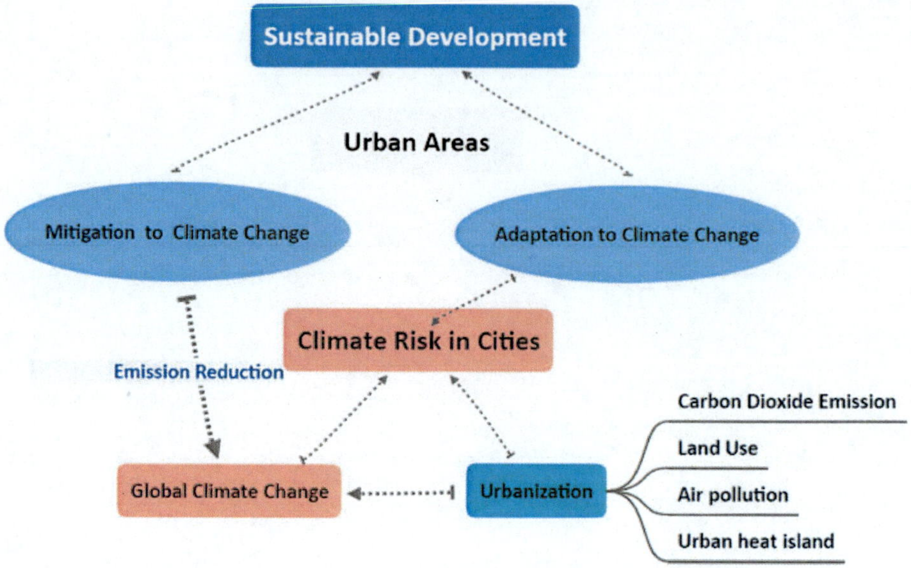

Fig. 1.5 Interactions between cities and climate change (drawn by the authors)

and risks, leading to a series of environmental problems, such as the urban heat island effect (UHIE), air pollution, urban flood, environmental, social, and economic loss. Active action is needed to address the challenges associated with global climate change and urban environmental degradation. Sustainable urban planning is one of the most important ways to mitigate threats and risks that may be caused by climate (Cheshmehzangi & Butters, 2017). Sustainable, safe, equitable, and resilient cities and communities need to be created in the face of global and local climate change challenges. The first is to identify the drivers of global climate change at the urban level. It is then vital to assess climate-induced risks that threaten cities to promote strategies and technologies (see Chap. 3). Besides, developing decision-making tools. Finally, designing government and industry guidelines, policies, and regulations to implement climate change mitigation and adaptation (as shown in Fig. 1.5).

1.4 Aim and Structure of the Book

In the beginning, this book reviews the changes and contributions that cities in different countries and regions have made in coping with climate change, mainly through literature review and analysis of historical data. The first two chapters mainly highlight the urgency of the impacts of climate change on cities and residents. Serving as the introduction, the first two chapters explore current progress on sustainable and climate-resilient development. As a complementary, the third chapter mainly introduces the co-benefits of carbon reductions, exploring various categories and areas of research for low carbon development (Cheshmehzangi, 2020b). As the next stage's background, Chaps. 4 and 5 explore more of the low carbon transitions globally and in China.

In the next core stage of the book, we mainly review China's sustainable development history and further analyse China's low carbon development according to some typical examples. In this regard, Chaps. 6 and 7 review the origin, process, and target plan of China's sustainable development while taking some achievements and attempts made in China as examples. China's 2030 and 2060 are taken as two major time nodes, and the process of carbon neutrality is divided into different stages. We delve into detailed discussions and viewpoints in Chaps. 8 and 9.

In the last part of the book, we assess China's sustainability performance in Chap. 10. Further discussions on carbon neutrality directions follow this. Therefore, we compare measures taken by different countries or cities to achieve carbon neutrality with China's progress. Chapter 11 provides a set of suggestions and paradigms to ensure our discussions lead to potential directions, policy changes, and practical suggestions. Our last chapter offers some concluding remarks on how China achieves or could achieve a green, low carbon, and high-quality development path. Some suggestions are put forward for China to achieve carbon neutrality in different periods. We believe lessons from China's pathways towards low carbon and climate-resilient plan could help other countries that are still yet to consider low carbon transitions. The findings from China's example are enlightening and hopefully helpful for future research.

In doing so, we summarise the book's aim and suggest sustainability transitions that are low carbon and climate-resilient against the current environmental crises and prolonged climate change challenges. We do not anticipate these transitions to be easy, but we expect to have the right tools and mindsets in achieving them promptly. Figure 1.6 summarises the framework and logic of the book.

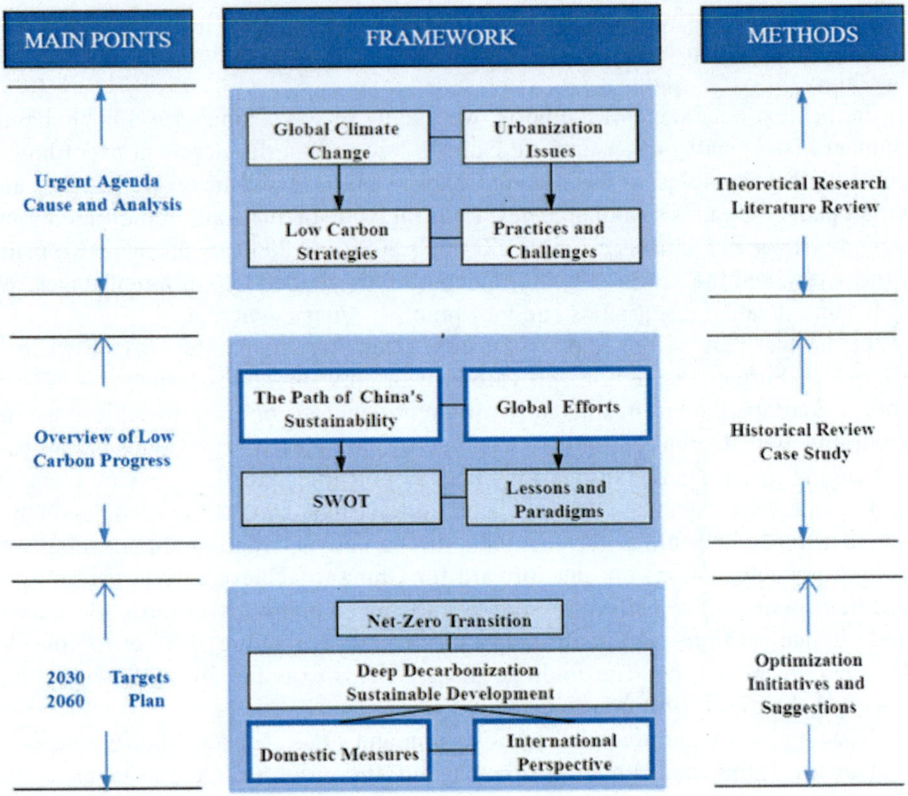

Fig. 1.6 The framework and logic of the book (drawn by the authors)

References

Banuri, T. (2009). climate change and sustainable development. *Natural Resources Forum, 33*(4), 257–258. https://doi.org/10.1111/j.1477-8947.2009.01270.x.

Butters, C., Cheshmehzangi, A., & Sassi, P. (2020). Cities, energy and climate: seven reasons to question the dense high-rise city. *Journal of Green Building, 15*(3), 197–214.

Chatterjee, N. (2011). Sustainable development and climate change. *International Journal of Environmental Studies, 68*(5), 756–758. https://doi.org/10.1080/00207233.2011.610122.

Cheshmehzangi, A. (2020a). The analysis of global warming patterns from 1970s to 2010s. *Atmospheric and Climate Sciences, 10*(3), 392–404.

Cheshmehzangi, A. (2020b). Low carbon transition at the township level: Feasibility study of environmental pollutants and sustainable energy planning. *International Journal of Sustainable Energy*, 1–27.

Cheshmehzangi, A., & Butters, C. (Eds.). (2017). *Designing cooler cities: Energy, cooling and urban form: The Asian perspective*. Palgrave Macmillan.

Cheshmehzangi, A., & Dawodu, A. (2019). The review of sustainable development goals (SDGs): People, perspective and planning. In *Sustainable urban development in the age of climate change* (pp. 133–156). Palgrave Macmillan.

Cheshmehzangi, A., Li, H., & Yang, X. (2018, October). Low carbon and climate resilient urban development in China. ADB Report, developed for the National Development Reform Committee (NDRC), Beijing, China.
Intergovernmental Panel on Climate Change (IPCC) (2014). *Fifth Assessment Report.* Retrieved December 17, 2020, from https://www.ipcc.ch/assessment-report/ar5/..
Lawn, P. (2016). *Resolving the climate change crisis: The ecological economics of climate change.* Springer. https://doi.org/10.1007/978-94-017-7502-1_2.
Mccarthy, M. P. , Best, M. J. , & Betts, R. A. (2010). Climate change in cities due to global warming and urban effects. *Geophysical Research Letters, 37*(9).
Munasinghe, M. (2001). Sustainomics, sustainable development and climate change. *Energy & Environment, 12*(5), 393–414. https://doi.org/10.1177/0958305X0101200501.
National Aeronautics and Space Administration (NASA). (2021a). *Climate change: How do we know?* Retrieved March 27, 2021. Available from https://climate.nasa.gov/evidence/.
National Aeronautics and Space Administration (NASA). (2021b). 2020 tired for warmest year on record, NASA Analysis. Available from https://www.nasa.gov/press-release/2020-tied-for-warmest-year-on-record-nasa-analysis-shows..
National Oceanic and Atmospheric Administration (NOAA). (2020). November and 2020 year to date rank 2nd hottest on record for globe. Retrieved December 17, 2020, from https://www.noaa.gov/news/november-and-2020-year-to-date-rank-2nd-hottest-on-record-for-globe
Organisation for Economic Co-operation and Development (OECD). (2020), *Environment at a Glance 2020.* Retrieved January 21, 2021 from https://doi.org/10.1787/4ea7d35f-en.
Soyez, K., & Graßl, H. (2008). *Climate change and technological options: Basic facts, evaluation and practical solutions.* Springer. https://doi.org/10.1007/978-3-211-78203-3_6.
Turner, R. K. (1995). Sustainable development and climate change. *Studies in Environmental Science, 65*, 55–66. https://doi.org/10.1016/S0166-1116(06)80194-9.
United Nations Environment Programme (UNEP). (2020). Emissions Gap Report 2020. Retrieved January 21, 2021 from https://www.unenvironment.org/emissions-gap-report-2020
U.S. Global Change Research Program (USGCRP). (2017). Fourth climate assessment. Retrieved December 17, 2020 from http://www.globalchange.gov/nca4.

Chapter 2
Cities and Climate-Resilient Development

2.1 Introduction to Impacts of Climate Change on Cities and Challenges Ahead

Cities provide not only jobs and economic activity but also essential social, environmental, and cultural services, which makes cities particularly vulnerable to climate change. Many related studies on household or urban water supply issues impact building infrastructure, coastal area management, and energy issues. Many detailed research studies have focused on the risk assessment of coastal cities and how to deal with rising sea levels. A few studies have also addressed other issues, including the impact(s) on tourism, cultural facilities, or biodiversity. (Hallegatte & Corfee-Morlot, 2011). Climate change affects ecosystems, water resources, food production, human settlements, and the frequency and scale of extreme weather events with significant consequences for human well-being and economic output (OECD, 2020). Climate change is occurring faster than expected, and GHG emissions could rise again (Araujo et al., 2007). We see no turning point in the current trends, and climate change issues have turned into complex situations of our time. As the recent reports state (ibid), the OECD countries continue to rely on fossil fuels for about 80% of their energy supply. This trend is impactful and may become more difficult to reverse in the long run.

Climate change poses challenges to urban systems in many ways. The first is the continuous warming of the climate (Cheshmehzangi & Dawodu, 2019). Hot days and hot nights are warmer and more frequent, which will increase the demand for water and various resources. As the discharge of sewage and waste increases, industrial demand and the energy consumption of urban operations will increase. Climate change increases the burden on urban systems and impacts on infrastructure, increasing ecological security risks (Cheshmehzangi & Butters, 2017). Secondly, the frequency of extreme catastrophic weather and climate events is likely to increase, which will greatly affect the economic and social development of cities and the lives of residents, and the construction of safety facilities is required to be improved. Higher requirements have been put forward for emergency measures and energy

A. Cheshmehzangi and H. Chen, *China's Sustainability Transitions*,
https://doi.org/10.1007/978-981-16-2621-0_2

supply guarantee measures in case of disasters. Besides, as the sea level continues to rise due to climate change, the problems of salt tide flooding and groundwater intrusion will become more serious, which will directly threaten coastal cities. The melting of glaciers and multi-year snow cover caused by climate change may increase the amount of surface runoff in downstream cities in the short term and even lead to flood disasters. However, cities based on snow and ice water will face the threat of water shortage and need to develop new water sources.

As climate change worsens, dangerous weather events are becoming more frequent or severe. People in cities and towns face the consequences, from heatwaves and wildfires to coastal storms and flooding. "Global Report on Human Settlements 2011: Cities and Climate Change" released by the UN-HABITAT put forward these impacts are a result of climate changes:

- Warmer and more frequent hot days and nights over most land areas;
- Fewer cold days and nights in many parts of the world;
- Frequency increases in warm spells/heat waves over most land areas;
- Increased frequency of heavy precipitation events over most areas;
- Increase in areas/regions affected by drought;
- Intense tropical cyclone activity in some parts of the world;
- Incidence of extreme high sea levels in some parts of the world.

Climate change is increasing the magnitude of many threats to urban areas. They are already being experienced because of rapid urbanisation, especially in the last few decades. With the development and manipulation of the environment in the industrial age, the effects of urbanisation and climate change have led to unprecedented negative impacts upon the quality of urban life, economic and social stability. Urban areas, with their high concentration of population, industries, and infrastructure, tend to face the most severe impacts of climate change. Although the innovation of strategies can control the GHG emissions or mitigation while coping mechanisms, disaster warning systems can reduce climate change impacts. Many cities have seen rapid and immense population growth in developing countries. The influx has been so significant, especially in countries of the developing world. Therefore, urban areas that are growing fastest are also those that are least equipped to deal with the threat of climate change and other environmental and socio-economic challenges. These areas often have profound deficits in governance, infrastructure, and economic and social equity. Climate change is an outcome of human-induced driving forces such as the combustion of fossil fuels and land-use changes, but with wide-ranging consequences for the planet and human settlements all over the world. The range of effects includes warming of sea-water, the melt of polar ice is accelerating a dangerous sea-level rise that threatens many coastal urban centres.

At the same time, the increasingly warm seas threaten the very existence of coral reef ecosystems around the world. With sea-level rise, urban areas along the coasts, particularly those in coastal zones, will be threatened with flooding, saltwater intrusion, and reductions in liveable space. Even in non-coastal areas, climate change can be very hazardous. Uncontrolled growth of urban centres into natural forest or brush areas that will dry out with increases in temperatures and the intensity and duration of droughts will see increases (Robinson et al., 2006). For instance, droughts in both coastal and non-coastal cities could disrupt urban water supplies and supplies of forest and agricultural products. The need for responses to an increased frequency of disasters will stress national economies even in developed countries, (also) creating much higher stress on the global economy. The challenges associated with the rapid pace of urbanisation will complicate responses to climate change.

IPCC claimed climate change as one of society's most significant threats in the coming decades (IPCC, 2018). As a result, there is a growing recognition that a broader response to climate change is needed, focusing on identifying processes and mechanisms for adaptation. The UK Climate Impacts Programme (UKCIP) shows that by 2050, our climate change will start to become apparent, and by 2080, there will be a significant change. There is a growing recognition that we need to prepare for the long-term consequences of climate change. This is especially significant for urban environments and infrastructure.

2.2 The Importance of Cities in Facing the Climate-Resilience Challenges

Emissions of greenhouse gases from urban activities disturb the radiative energy balance of the earth-atmosphere system, leading to temperature changes and other disruptions of the earth's climate. Most emissions stem from energy use in transport, industry, agriculture, and land-use. The other side of the coin is that cities will also offer many opportunities to develop responses in both mitigation and adaptation strategies to deal with climate change. The populations, enterprises, and authorities of cities will be fundamental players in developing these strategies. In this way, climate change itself will offer opportunities, or it will force cities and humanity, in general, to improve global, national, and urban governance to foster the realisation of sustainable development.

The IPCC approved a proposal for the co-sponsored International Conference on Climate Change and Cities, held in Edmonton, Canada, in 2018 (IPCC, 2018). The "*Global Research and Action Agenda on Cities and Climate Change Science*" were co-produced during the Conference, and the structure of it is illustrated in Fig. 4. The Conference highlighted a range of crosscutting knowledge gaps, including System Approach, City-level Models & Data, Scale, and Governance. Some key topical knowledge gaps needed to be addressed, such as Informality, Urban Planning & Design, Sustainable Consumption & Production, Built & Green/Blue Infrastructure, and so on (IPCC, 2018) (see Fig. 1.6) (Fig. 2.1).

The United Nations estimates that by 2050, the world's urban population will reach 6.3 billion. The role of cities in addressing climate change is critical within the context of urban population expansion. Although Climate change exacerbates the pressures of rapid population growth and sprawl, poverty, and pollution, cities have the tools and resources that help tackle climate change challenges. The number of

Fig. 2.1 Climate adaptation and mitigation in cities (drawn by the authors, data is extracted from IPCC, 2018)

initiatives and interventions in cities that seek to address climate change appears to be rapidly proliferating. Some of them may relate to eco-developments, new technologies, specific policies, community-based initiatives, corporate buildings, infrastructure renewal programmes, or the like. Climate change is increasingly attaching itself to the development, repair, and maintenance of the city (Bulkeley & Castán Broto, 2013). Harlan and Ruddell (2011) summarised the current status of urban climate change, the impact of two urban climate disasters, namely temperature rise and air pollution, public health, and the risks of different urban populations. They sorted out the main contents of the current urban climate change risk management plan, environmental impact, and potential health benefits. Bai et al. (2018) proposed six key points of urban and climate change research expand observations, understand climate Interactions, study informal Settlements, harness currents technologies, support transformation, recognise global sustainability context. According to local differences, cities' climate-related policies can be divided into mitigation policies and adaptation policies, or three types of plans—Be made independently, in accordance with national goals, in association with international climate networks (Reckien et al., 2018).

Based on different studies, it can be concluded that cities' measures in the face of climate change can be divided into urban design, built environment, infrastructure, transportation, and logistics. Precisely, they follow some basic principles, such as balancing benefits and risks in terms of energy demand, maintaining a sustainable greenhouse gas emission footprint, addressing both short-term and long-term problems with mitigation actions, and supporting multi-industry and multi-sector response policies.

2.3 Decarbonisation Progress and Towards Low Carbon Economy

2.3.1 *Statements by Leading Organisations*

The number of transnational municipal networks engaged with the climate change issue has increased. In contrast, members like UN-Habitat, WWF, Action Aid, Transition Towns, HSBC, the Clinton Climate Foundation, the Rockefeller Foundation, and the World Bank have sought to mobilise action to 'Low Carbon Economy'. Many statements have been made by the world's leading organisations in history.

In June 1992, the United Nations Conference on Environment and Development (UNCED) held in Rio de Janeiro put forward an international environmental treaty, namely the United Nations Framework Convention on Climate Change (UNFCCC or FCCC), which came into force in 1994. Since 1995, the Parties to the United Nations Framework Convention on Climate Change (UNFCCC) have convened the Conference of the Parties (COP) annually to assess progress in combating climate change. The Kyoto Protocol was signed in 1997. This was the first time in human history that greenhouse gas emissions have been limited in the form of regulations. In 2008, the United Nations Climate Change Conference was held in Bangkok, Thailand, which was the first time since the Bali Roadmap was agreed in 2007. Next year, the G8 Summit was held in July. Members have issued a statement to work with other countries to reduce global greenhouse gas emissions by at least 50 percent by 2050. The UN Climate Change Summit will be held in September 2009, and it was expected that an effective agreement would be reached. On December the 15th, 2009, the Conference of UNFCCC was held in Copenhagen, which made arrangements for global greenhouse gas reductions after 2012. Then the 2010 Cancun agreement also stipulated that global warming should be kept below 2.0 °C (3.6 degrees Fahrenheit) above pre-industrial levels. The 17th Conference of the Parties (COP 17) in 2011 and the 7th Conference of the Parties to the Kyoto Protocol (COP 7) were held in Durban, South Africa. The 19th Conference of the Parties (COP 19), held in Warsaw, Poland, in 2013, adopted the Durban Platform. These reoccurring international events signify the scale of crises and their impacts on our everyday life and the future.

At the UN Sustainable Development Summit held in New York in September 2015, 17 Sustainable Development Goals (SGDs) were formally introduced and adopted. Among those Goals, the 13th goal, namely 'Climate Action', was proposed to strengthen resilience and adaptive capacity to climate-related hazards and natural disasters in all countries. The agreement adopted at the Paris Climate Change Conference in December 2015 came into force in 2016. Commonly known as the Paris talks, the event helped push for a common goal against climate change impacts. This led to a wide-scale agreement, and with its vicissitudes (or some political ups and downs), it remains in place. The Paris Agreement is the third crucial legal text of the human response to climate change, following the United Nations Framework Convention on Climate Change and the Kyoto Protocol. The Paris Agreement calls for limiting the increase in global average temperature to less than 2 °C above pre-industrial levels over the course of this century and for efforts to limit the rise in temperature to 1.5 °C above pre-industrial levels. It requires countries to participate in the fight against climate change in the form of "Nationally Determined Contributions (NDC)".

World Meteorological Organisation (WMO) issued a statement on the State of the Global Climate of 2019. The year 2019 was the second warmest year in the instrumental record. Besides, 2015–2019 are the five warmest years on record, and 2010–2019 is the warmest decade on record. Each successive decade has been warmer than any preceding decade since 1850. Despite the temporary drop in CO_2 emissions caused by the COVID-19 pandemic, the world is still on track to rise by more than 3 °C by the end of the century, which is well beyond the Paris Agreement's goal of keeping global warming below 2 °C. About a quarter of the G20 countries have a fixed share of national spending (no more than 3% of GDP) that is explicitly devoted to low carbon measures. The growing number of countries committing to "Net Zero Emissions" by the middle of the century has emerged as a critical climate policy development for 2020. The 126 countries accounting for 51% of global greenhouse gas emissions by 2020 have announced or are considering achieving "Net Zero Emissions Targets". More countries need to develop long-term strategies in line with the Paris Agreement, and the new and updated Intended NDC needs to be in line with the "Net Zero Emissions Target" (UNEP, 2020).

2.3.2 Urban Responses to Climate Change

The earliest wave of urban responses to climate was in the early 1990s, predominantly in North America and Europe. There were three international networks. Firstly, the International Council launched some CO_2 Reduction Programme, which was set up for Local Environmental Initiatives. Next was the Climate Alliance, which was founded in 1990. The main aim was to reduce emissions and protect the rainforests. It was notable that this was an alliance between European cities and indigenous peoples. Energy cites formed in 1990 involved a European Union project and several European countries (Bulkeley, 2010).

Since the beginning of the twenty-first century, coalitions of cities actively addressing climate change have further emerged. The World Mayors' Council on Climate Change (WMCCC) was established in 2005. And that led to the launch of the C40 Urban Climate Leadership Group (The map was shown as demonstrated in Fig. 1.7), which was made up of many different cities worldwide. It aimed to reduce GHG emissions through a range of energy projects (Rosenzweig et al., 2010) (Fig. 2.2).

Many cities have taken positive actions from both mitigation and adaptation perspectives in various ways to enhance urban governance capacity to cope with the impacts of climate change. The mitigation strategies of cities to deal with climate change mainly reduce the proportion of greenhouse gases in the atmosphere. Some cities reduce greenhouse gas emissions by compiling emission inventories and reducing emissions from key industries and sectors. The 2 °C project in Chicago, for

Fig. 2.2 The C40 cities climate leadership group of large cities (drawn by the authors, data extracted from Rosenzweig et al, 2010)

example, has compiled a carbon mitigation guidelines manual. In Berlin, Germany, emissions inventories focus on power and transport systems. New York has also proposed its long-term growth and sustainability plan, known as the 2030 Plan, to reduce greenhouse gas emissions by 30 percent from 2005 levels over the next 20 years. It also included mandatory energy audits of cities and commercial buildings. Another side is to focus on carbon reduction in key industries. International advanced cities usually focused mitigation actions on energy, construction, transportation, waste, and urban development to achieve industry-level mitigation. The energy scheme of London stipulated that all new projects must be integrated into a distributed cogeneration system. In the construction sector, Berlin has enacted regulations to promote energy efficiency in buildings and developed important tools with international impact.

The adaptation strategy of cities to climate change was to strengthen the emergency management of climate disaster. In general, international cities mainly focused on the reconstruction and improvement of urban infrastructure, research on urban problems and disasters, and the definition of urban vulnerability and risk. Besides, there were efforts to enhance citizens' ability to adapt to climate change, attach importance to the communication of media and information technology, and establish rapid response mechanisms to improve citizens' ability to help themselves and help each other when disasters occur.

2.4 Conclusions

In this chapter, we briefly discussed how to improve the ability of urban climate change response and adaptation. This starts with identifying the gap between existing urban systems and the needs of climate change and how best to bridge that gap. Second, environmental monitoring and adaptation planning should be developed in response to climate change (Kaplan, 2014). Finally, promote the participation of organisations and individuals in the decision-making and implementation of relevant measures. Climate change will become an important factor restricting urban development and upgrading in the future. In the process of urban planning and environmental and climate governance, it is necessary to investigate the impact of extreme climate events and change strategies. It also needs to integrate urban risk, vulnerability, and resilience strategies and paths (Xiao et al., 2011). Besides, it is particularly important to study the correlation between infrastructure construction and the ecological environment. Progress on the proposed climate response measures, which include not only short-term climate challenges such as extreme temperatures, heavy rainfall, and coastal flooding, must also be carried out in planned phases and extended to non-urban areas.

References

Bai, X., Dawson, R. J., Urge-Vorsatz, D., Delgado, G. C., Barau, A. S., & Dhakal, S., et al. (2018). Six research priorities for cities and climate change. *Nature, 555*(7694), 23.

Bulkeley, H. (2010). Cities and the governing of climate change. *Annual Review of Environment and Resources, 35*(1), 361–375(15)

Bulkeley, H., & Castán Broto, V. (2013). Government by experiment? Global cities and the governing of climate change. *Transactions of the Institute of British Geographers, 38.*

Cheshmehzangi, A., & Butters, C. (Eds.). (2017). *Designing cooler cities: Energy, cooling and urban form: The Asian perspective.* Palgrave Macmillan.

Cheshmehzangi, A., & Dawodu, A. (2019). *Sustainable urban development in the age of climate change—people: The cure or curse.* Palgrave Macmillan.

de Araujo, M. S. M., de Campos, C. P., & Rosa, L. P. (2007). GHG historical contribution by sectors, sustainable development and equity. *Renewable & Sustainable Energy Reviews, 11*(5), 988–997.

Hallegatte, S., & Corfee-Morlot, J. (2011). Understanding climate change impacts, vulnerability and adaptation at city scale: An introduction. *Climatic Change, 104*(1), 1–12.

Harlan, S. L., & Ruddell, D. M. (2011). Climate change and health in cities: Impacts of heat and air pollution and potential co-benefits from mitigation and adaptation. *Current Opinion in Environmental Sustainability, 3*(3), 126–134.

Intergovernmental Panel on Climate Change (IPCC). (2018). *Global research and action agenda on cities and climate change science.* Retrieved January 21, 2021 from https://www.ipcc.ch/site/assets/uploads/2019/07/Research-Agenda-Aug-10_Final_Long-version.pdf.

Kaplan, M. (2014). Climate-resilient development—participatory solutions from developing countries. *European Journal of Development Research, 26*(5), 922–924.

Organisation for economic co-operation and development (OECD). (2020), *Environment at a Glance 2020.* Retrieved January 21, 2021 from https://doi.org/10.1787/4ea7d35f-en.

Reckien, D., Salvia, M., Heidrich, O., Church, J. M., Pietrapertosa, F., De Gregorio-Hurtado, S., & Dawson, R. (2018). How are cities planning to respond to climate change? Assessment of local climate plans from 885 cities in the EU-28. *Journal of Cleaner Production, 191*, 207–219.

Robinson, J., Bradley, M., Busby, P., Connor, D., Murray, A., Sampson, B., & Soper, W. (2006). Climate change and sustainable development: Realizing the opportunity. *Ambio A Journal of the Human Environment, 35*(1), 2–8.

Rosenzweig, C., Solecki, W., Hammer, S. A., & Mehrotra, S. (2010). Cities lead the way in climate-change action. *Nature, 467*(7318), 909–911.

United Nations Environment Programme (UNEP). (2020). *Emissions Gap Report 2020.* Retrieved January 21, 2021 from https://www.unenvironment.org/emissions-gap-report-2020.

Xiao, L., Li, X., & Wang, R. (2011). Integrating climate change adaptation and mitigation into sustainable development planning for Lijiang city. *International Journal of Sustainable Development & World Ecology, 18*(6), 515–522.

Chapter 3
Summarising Key Carbon Reduction Co-benefits

3.1 The Significance of Reducing Carbon for the Health of People and the Planet

Reducing carbon emissions is widely recognised as a foremost driver for healthy people and the planet (Foley 2010; Paterson and Stripple 2010; Spence 2014; Asrar et al. 2019; Ekins and Gupta 2019). This is particularly discussed in correlation with climate change issues and the broader ecological calamity (Cheshmehzangi and Dawodu 2019; Harvie and Guarneri 2020), as well as the sustainable development goals (Griggs et al. 2013), and health benefits (Milner et al. 2020). A major part of carbon reduction strategies is to pursue air pollutant co-benefits (Dong et al. 2015), leading to mitigation strategies at multiple spatial levels of the built and natural environments. These strategies are vital to the health of people and the planet, suggesting the feasibility of decarbonisation processes and reduction of environmental pollutants in the built environments (Cheshmehzangi 2020a, b). Cleaner air is the main aspect associated with air quality, including a wide range of health co-benefits (Driscoll et al. 2015). It overlaps with other aspects that suggest the importance of carbon reduction co-benefits for healthy living environments.

As highlighted by the Intergovernmental Panel on Climate Change (IPCC) (n.d.), global warming reduces global economic growth by 0.2–2.0% per annum. It poses severe threats to the cryosphere, ecology, water, food, safety, and human health of cities and significant projects (also see IPCC 2018; NOAA 2020). The threats are so prevalent that they make our attempts for adaptation strategies minimal and ineffective. The concerns come from increasing global warming and carbon emissions (UNEP 2020), particularly the latter that influences the former. As West et al. (2013) suggest, the future ozone reduction is primarily due to co-emitted air pollutants, estimated at a large share of 89% by 2100. The co-benefits, therefore, help to consider air pollution reduction and human health improvement (Kim et al. 2020) and towards carbon offsetting (MacKerron et al. 2009), reduction goals in other sectors (Jiang et al. 2013), and low carbon scenarios for the development of any kind (Dhar and Shukla 2015; Cheshmehzangi 2020c).

A. Cheshmehzangi and H. Chen, *China's Sustainability Transitions*,
https://doi.org/10.1007/978-981-16-2621-0_3

Deng et al.'s (2018) research findings are instrumental as they highlight the most up-to-date research streams of carbon reduction co-benefits. These research streams are also vital for achieving sustainable and low carbon pathways in future development. Their findings are summarised below (ibid, p. 1):

> …the co-benefits from GHG mitigation that have received the largest attention of researchers are impacts on ecosystems, economic activity, health, air pollution, and resource efficiency. The co-benefits that have received the least attention include the impacts on conflict and disaster resilience, poverty alleviation (or exacerbation), energy security, technological spillovers and innovation, and food security. Most research has investigated co-benefits from GHG mitigation in the agriculture, forestry and other land use (AFOLU), electricity, transport, and residential sectors, with the industrial sector being the subject of significantly less research. The largest number of co-benefits publications provide analysis at a global level, with relatively few studies providing local (city) level analysis or studying co-benefits in Oceanian or African contexts. Finally, science and engineering methods, in contrast to economic or social science methods, are the methods most commonly employed in co-benefits papers.

To follow on these valuable findings, we summarise carbon reduction co-benefits into six categories of (1) air quality, (2) health and wellbeing, (3) energy conservation, (4) energy efficiency, (5) biodiversity, and (6) resource efficiency (Fig. 3.1). In doing so, we recognise the central aspect of decarbonisation processes (Cheshmehzangi 2016a; Cheshmehzangi et al. 2021; Gallagher and Holloway 2020), which also suggest environmental co-benefits (Luderer et al. 2019), sector-based carbon reduction cases (Creutzig et al. 2012; Dhar et al. 2017; Ma et al. 2016; Peng et al. 2018; Teng and Jotzo 2014), and improvements towards low carbon pathways (Li

Fig. 3.1 Six suggested categories of carbon reduction co-benefits

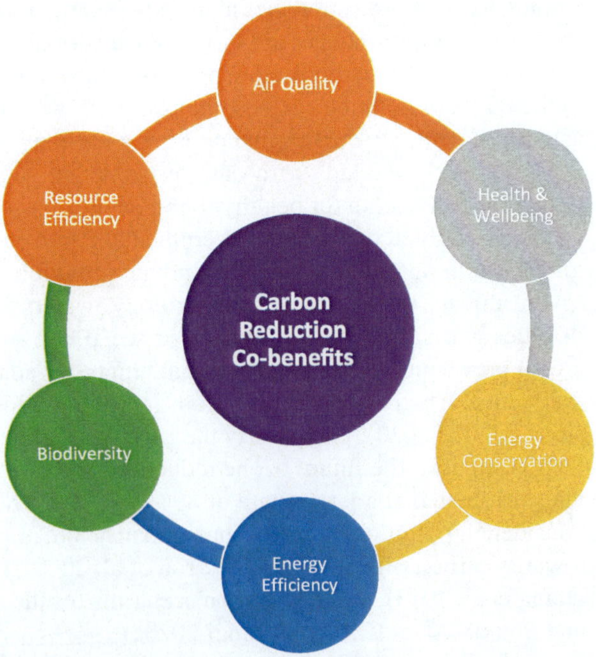

et al. 2019) or even low carbon transitions (Cheshmehzangi et al. 2018; Sovacool et al. 2019). To follow up, we scrutinise these six categories and demonstrate their importance in the main argument of healthier people and the planet. The following six sub-sections include what we believe are the core carbon reduction co-benefits, especially in dealing with the decarbonisation of the built environments.

3.2 Achieving a Healthy Air Quality

Healthy air quality is perhaps the main co-benefits of carbon reductions (Thompson et al. 2014). It is primarily associated with carbon pricing issues (Li et al. 2018; Parry et al. 2014), short-term and long-term reduction strategies, and quantifying carbon emissions (i.e., through modeling methods). The reduction of greenhouse gas (GHG) emissions implies huge contributions to future air quality and human health (West et al. 2013), something that could represent both emission controls (Anenberg et al. 2012; Harmsen et al. 2020) and the development of standards (Driscoll et al. 2015). Such methods' implications lead to better consideration of incorporating air quality co-benefits into climate change policymaking (Nemet et al. 2010) and context-specific examples of carbon policies (Thomson et al. 2016). The implications could be on individual cities or regions and national systems (Chang et al. 2020), including reducing energy-related demands, trade, and businesses, and achieving cleaner air productions. In reaching the former, we could benefit from having a much healthier society, which implies larger scaler health co benefits (Gao et al. 2018), which should be considered throughout the process. Some could happen at the micro level (Butters et al. 2020; Deng and Cheshmehzangi 2018), but some involve macro-level assessment and implications (Jensen et al. 2013), which are ultimately the foundation of achieving healthier air quality measures.

3.3 Enhancing Human Health and Wellbeing

The other overarching category is human health and wellbeing. Associated with environmental co-benefits (Barrett et al. 2016; Cohen and Kantenbacher, 2020; Rissel 2009), we see many implications for healthy living, such as public health co-benefits (Jack and Kinney 2010; Liu and Gao 2020) and wellbeing (Maccagnan et al. 2019). Similar to what West et al. (2013) highlight, there are many environmental co-benefits related to carbon reductions, which suggest the importance of carbon limitations, industrial restructuring, and the implementation of climate policy (Rypdal et al. 2007). Central to other co-benefits, human health and wellbeing are crucial in achieving a successful low carbon transition. Unlike the climate change adaptation plan (Tonmoy et al. 2020), the carbon reduction co-benefits related to health and wellbeing are quite comprehensive. This also includes other factors such as physical health and happiness and the relationship between the health-income paradox and satisfaction levels (Fanning and O'Neill 2019). These are often seen as part of developing policy interventions (Howden-Chapman et al. 2020) as well as secondary areas, such as food security (Mok et al. 2020). The example of food and water security, also embedded in the United Nations' Sustainable Development Goals (SDGs), are important areas that relate directly to societies' health and wellbeing. They open up opportunities for us to consider future challenges (Bahar et al. 2020) under the shadow of combined pollution and warming calamities. To date, there are not many studies that study the nexus between carbon reduction co-benefits and other health-related factors, but this area of research is proliferating. On food security, for instance, we could look into a wide range of factors beyond just air or water quality, but also soil contamination, food scarcity, croplands, greenfield qualities, etc. To enhance human health and wellbeing, we ought to consider the diversity of carbon reduction co-benefits, and particularly those that empower communities, promote innovations, and push for promising policies and governmental interventions.

3.4 Optimising Energy Conservation and Energy Security

The nexus between energy conservation and carbon reduction (Jiang et al. 2013) is studied extensively in the built environment sector, especially at the building level (Balaban and de Oliveira, 2017). Design of energy-conservation and emission-reduction plans (Li 2020) includes energy security matters directly related to carbon reduction co-benefits. Thus, the combined energy conservation and energy security factors play a significant part in the development of low carbon experimentation (Cheshmehzangi et al. 2018; Lo and Castán Broto 2019; Raven et al. 2019) and the development of energy-related measures, such as in the air pollution prevention and control action plan (Lu et al. 2019). From household energy projects (Karhunmaa 2016) to larger-scale industrial restructuring developments (Cheshmehzangi 2020c), the need for optimising energy conservation and energy security is evident. A broader understanding of this co-benefit category is related to carbon management mechanisms and tools (Milner et al. 2012), leading the way towards low carbon cities, mitigation strategies, and decarbonisation opportunities. In these areas, the energy-related research includes multiple energy security indicators (Lin and Yousaf Raza 2020), as well as other factors, such as energy equity and environmental sustainability measures (Fu et al. 2021). The existing studies also overlap renewable energy research and low carbon emission resources (He et al. 2020), carbon pricing (Galinis et al. 2020), and energy efficiency improvements (Trotta 2020). All these directly or indirectly address the overarching topics of energy conservation and energy security.

3.5 Augmenting Energy Efficiency and Reducing Energy Use

From low carbon transport (Dhar and Shukla 2015) to specific policy development examples (Howden-Chapman and Chapman 2012), we see many examples of cross-sector opportunities for actions against the increase of carbon emissions (Ramaswami et al. 2017). The reduction in energy demand could lead to larger level co-benefits (Shrestha and Pradhan 2010), which also suggests co-benefits through resource recovery (Menikpura et al. 2013), such as energy efficiency in water-energy nexus (Zhou et al. 2018). In particular, between sectors, the trade-offs that exist could reflect on adverse effects and create an adequate system of mitigation actions (Bustamante et al. 2014). Some of the energy efficiency studies focus on intensities of production and consumption, specifically CO_2 intensity reduction and air pollution abatement (Zhang et al. 2014, 2015). Some reductions could reduce production rates and intensities and reduce the embodied energy of building construction (Cheshmehzangi and Butters, 2016, 2017). In reducing embodied energy, in particular, many other benefits could help ease the path towards low carbon and low-cost solutions. For energy use, the same applies for operational energy, ranging from reductions in household consumption (Levy et al. 2016) to industrial restructuring opportunities (Cheshmehzangi 2020b) and towards low carbon and cleaner productions. With possibilities to reduce electrical energy demand (Selvakkumaran and Limmeechokchai 2013), the augmentation of efficiency and reduction of energy use could play a significant part in achieving the low carbon targets.

3.6 Supporting Biodiversity Through Reduced Environmental Crises

Much of the research in this area focus on climate change and biodiversity losses (Phelps et al. 2012), known to be leading environmental crises of our time. The role of conservation policy and action has proven to be an essential attempt to increase carbon storage capacity and enhance co-benefits for biodiversity (Soto-Navarro et al. 2020). The higher-level forestation initiatives provide augmented co-benefits to our biodiversity (Gilroy et al. 2014; Deere et al. 2018; Buotte et al. 2020; Cheshmehzangi et al. 2021) and the other living habitats, including the cities. Much of the recent in this area is focused on sustainable forest management in biodiversity conservation and carbon sequestration (Imai et al. 2009) and supporting biodiversity and ecosystem services (Bertzky et al. 2010; Smith et al. 2019). As a significant co-benefit of carbon reductions, reducing the environmental crises could lead to greener living environments and more sustainable developments. As discussed by Jiang et al. (2013), GHG reductions' examples respond to larger-scale CO_2 mitigation opportunities (Dong et al. 2015), particularly by considering the environmental benefits and environmental sustainability measures. Besides, extensive social benefits could reduce the existing inequalities in our communities around the globe (Wittman and Caron 2009). For instance, these include helping to reduce environmental and natural disasters, moving towards enhancing disaster resilience, food security, water security, etc. By supporting biodiversity through reduced environmental crises, we could create a better cycle of ecosystem services and ecological qualities in both natural and built environments.

3.7 Maintaining Resource Efficiency

The combination of resource efficiency and climate policy could lead to positive solutions in terms of reductions in production and consumption patterns (Ekins et al. 2016). A remarkable example is material efficiency strategies (Hertwich et al. 2020), particularly for low carbon pathways and GHG emission reductions. In the areas of healthy buildings and sustainable cities, as Balaban and de Oliveira (2017) argue, a lot can be done to systematically enhance our resource efficiency and reduce the pollution levels/ratios (Bringezu et al. 2017). One of the examples is "*decoupling economic activity and human well-being from resource use*" (ibid, p. 8), which could be done by enhanced resource efficiency and achieving the SDGs. Another example is the circular economy and its ideals that focus on socio-economic benefits and long term sustainable measures. All these examples help to achieve a better set of pollution control strategies, "*by lowering the amount of resources used, the amount of related emissions and impacts can also be reduced, and many of them at the same time*" (ibid, p. 9). This also helps to transform the economy through sustainable and low carbon transitions, which has become commonly advocated by many governments; or at least has become a common axiom to combat climate change and environmental crises. To reduce resource use and maintain it, we ought to consider global action plans that suggest proper sustainability transitions. Methods of integrated resource management, and action plans against reducing resource use, are no longer unique examples but are simply strategies that make sense if we wish to move towards a low carbon future.

3.8 Conclusions

In this chapter, we explored the importance of carbon reduction from six different perspectives. Reducing carbon footprint means low energy, low consumption, and low cost. A low carbon lifestyle is not only beneficial to people's health but also can greatly improve the ecological environment and reduce the loss of hazards (Markandya et al. 2009; Sovacool et al. 2020). Carbon reduction is a relatively new direction of sustainable development in the world, which reflects the concern of human beings for the future due to climate change. Excessive carbon dioxide emissions that cause climate change are caused by human production and consumption, so the key to solving the problem lies in human beings themselves (Tan et al. 2014; Cheshmehzangi and Dawodu 2019). Today, with the increasing use of energy, the implementation of energy conservation and carbon reduction is the way towards the new century.

References

Anenberg, S. C., Schwartz, J., Shindell, D., Amann, M., Faluvegi, G., Klimont, Z., Janssens-Maenhout, G., Pozzoli, L., Van Dingenen, R., Vignati, E., & Emberson, L. (2012). Global air quality and health co-benefits of mitigating near-term climate change through methane and black carbon emission controls. *Environmental Health Perspectives, 120*(6), 831–839.

Asrar, G. R., Lucas, P., van Vuuren, D., Pereira, L., Vervoort, J., & Bhargava, R. (2019). Outlooks in geo-6-global environment outlook (geo-6): Healthy planet, healthy people chapter 19. *Global Environment Outlook (GEO-6): Healthy Planet, Healthy People.*

Bahar, N. H. A., Lo, M., Sanjaya, M., Van Vianen, J., Alexander, P., Ickowitz, A., & Sunderland, T. (2020). Meeting the food security challenge for nine billion people in 2050: What impact on forests? *Global Environmental Change, 62*, 102056.

Balaban, O., & de Oliveira, J. A. P. (2017). Sustainable buildings for healthier cities: Assessing the co-benefits of green buildings in Japan. *Journal of Cleaner Production, 163*, S68–S78.

Barrett, B., Grabow, M., Middlecamp, C., Mooney, M., Checovich, M. M., Converse, A. K., Gillespie, B., & Yates, J. (2016). Mindful climate action: Health and environmental co-benefits from mindfulness-based behavioral training. *Sustainability, 8*(10), 1040.

Bertzky, M., Ravilious, C., Araujo Navas, A. L., Kapos, V., Carrión, D., Chíu, M., & Dickson, B. (2010). *Carbon, biodiversity and ecosystem services: Exploring co-benefits—Ecuador.* Cambridge: UNEP World Conservation Monitoring Centre.

Bringezu, S., Ramaswami, A., Schandl, H., O'brien, M., Pelton, R. E. O., Acquatella, J., Ayuk, E., Chiu, A. S. F., Flanegin, R. K. Fry, J., Giljum, S., Hashimoto, S., Hellweg, S., Hosking, K., Hu, Y., Lenzen, M., Lieber, M., Lutter, S., & Zivy, R. (2017). Assessing global resource use: A system approach to resource efficiency and pollution reduction. *IRP Reports.* 104 p. Available from: https://stg-wedocs.unep.org/bitstream/handle/20.500.11822/27432/resource_use.pdf?sequence=1&isAllowed=y.

Buotte, P. C., Law, B. E., Ripple, W. J., & Berner, L. T. (2020). Carbon sequestration and biodiversity co-benefits of preserving forests in the western United States. *Ecological Applications, 30*(2), e02039.

Bustamante, M., Robledo-Abad, C., Harper, R., Mbow, C., Ravindranat, N. H., Sperling, F., de Siqueira Pinto, A., & Smith, P. (2014) Co-benefits, trade-offs, barriers and policies for greenhouse gas mitigation in the agriculture, forestry and other land use (AFOLU) sector. *Global Change Biology, 20*(10), 3270–3290.

Butters, C., Cheshmehzangi, A., & Sassi, P. (2020). Cities, energy and climate: Seven reasons to question the dense high-rise city. *Journal of Green Building, 15*(3), 197–214.

Chang, S., Yang, X., Zheng, H., Wang, S., & Zhang, X. (2020). Air quality and health co-benefits of China's national emission trading system. *Applied Energy, 261*, 114226.

Cheshmehzangi, A. (2016a). China's new-type urbanisation plan (NUP) and the foreseeing challenges for decarbonization of cities: A review. *Energy Procedia, 104*, 146–152.

Cheshmehzangi, A. (2016b). Feasibility study of Songao's low carbon town planning, China. *Energy Procedia, 88*, 313–320.

Cheshmehzangi, A. (2020a). The analysis of global warming patterns from 1970s to 2010s. *Atmospheric and Climate Sciences, 10*(3), 392–404.

Cheshmehzangi, A. (2020b). Low carbon transition at the township level: Feasibility study of environmental pollutants and sustainable energy planning. *International Journal of Sustainable Energy*, 1–27.

Cheshmehzangi, A. (2020c). Towards urban energy management: A brief review of China's recent transition in urban energy demand. *Low Carbon Economy, 11*(1), 1–24.

Cheshmehzangi, A., & Butters, C. (2016). Sustainable living and urban density: The choices are wide open. *Energy Procedia, 88*, 63–70.

Cheshmehzangi, A., & Butters, C. (Eds.). (2017). *Designing cooler cities: Energy, cooling and urban form: The Asian perspective*. Palgrave Macmillan.

Cheshmehzangi, A., & Dawodu, A. (2019). The review of sustainable development goals (SDGs): People, perspective and planning. In *Sustainable urban development in the age of climate change* (pp. 133–156). Singapore: Palgrave Macmillan.

Cheshmehzangi, A., Dawodu, A., & Sharifi, A. (2021). *Sustainable urbanism in China*. Routledge.

Cheshmehzangi, A., Xie, L., & Tan-Mullins, M. (2018). The role of international actors in low-carbon transitions of Shenzhen's International Low Carbon City in China. *Cities, 74*, 64–74.

Cohen, S. A., & Kantenbacher, J. (2020). Flying less: Personal health and environmental co-benefits. *Journal of Sustainable Tourism, 28*(2), 361–376.

Creutzig, F., Mühlhoff, R., & Römer, J. (2012). Decarbonizing urban transport in European cities: Four cases show possibly high co-benefits. *Environmental Research Letters, 7*(4), 044042.

Deere, N. J., Guillera-Arroita, G., Baking, E. L., Bernard, H., Pfeifer, M., Reynolds, G., Wearn, O. R., Davies, Z. G., Struebig, M. J. (2018). High carbon stock forests provide co-benefits for tropical biodiversity. *Journal of Applied Ecology, 55*(2), 997–1008.

Deng, H. M., Liang, Q. M., Liu, L. J., & Anadon, L. D. (2018). Co-benefits of greenhouse gas mitigation: A review and classification by type, mitigation sector, and geography. *Environmental Research Letters, 12*(12), 123001.

Deng, W., & Cheshmehzangi, A. (2018). *Eco-development in China: Cities, communities and buildings*. Palgrave Macmillan.

Dhar, S., & Shukla, P. R. (2015). Low carbon scenarios for transport in India: Co-benefits analysis. *Energy Policy, 81*, 186–198.

Dhar, S., Pathak, M., & Shukla, P. R. (2017). Electric vehicles and India's low carbon passenger transport: A long-term co-benefits assessment. *Journal of Cleaner Production, 146*, 139–148.

Dong, H., Dai, H., Dong, L., Fujita, T., Geng, Y., Klimont, Z., Inoue, T., Bunya, S., Fujii, M., Masui, T. (2015). Pursuing air pollutant co-benefits of CO_2 mitigation in China: A provincial leveled analysis. *Applied Energy, 144*, 165–174.

Driscoll, C. T., Buonocore, J. J., Levy, J. I., Lambert, K. F., Burtraw, D., Reid, S. B., Fakhraei, H., & Schwartz, J. (2015). US power plant carbon standards and clean air and health co-benefits. *Nature Climate Change, 5*(6), 535–540.

Ekins, P., & Gupta, J. (2019). Perspective: A healthy planet for healthy people. *Global Sustainability*, 2.

Ekins, P., Hughes, N., Brigenzu, S., Arden Clark, C., Fischer-Kowalski, M., Graedel, T., Hajer, M., Hashimoto, S., Haield-Dodds, S., Havlík, P., Ingram, J. S. I., Kruit, K., Milligan, B. M., Moriguchi, Y., Nasr, N., Newth, D., Obersteiner, M., Ramaswami, A., & Westhowk, H. (2016). Resource efficiency: Potential and economic implications.

Fanning, A. L., & O'Neill, D. W. (2019). The wellbeing-consumption paradox: Happiness, health, income, and carbon emissions in growing versus non-growing economies. *Journal of Cleaner Production., 212*, 810–821.

Foley, J. (2010). Boundaries for healthy planet. *Scientific American, 302*(4), 54–57.

Fu, F. Y., Alharthi, M., Bhatti, Z., Sun, L., Rasul, F., Hanif, I., & Iqbal, W. (2021). The dynamic role of energy security, energy equity and environmental sustainability in the dilemma of emission reduction and economic growth. *Journal of Environmental Management, 280*, 111828. https://doi.org/10.1016/j.jenvman.2020.111828.

Galinis, A., Martišauskas, L., Jääskeläinen, J., Olkkonen, V., Syri, S., Avgerinopoulos, G., & Lekavičius, V. (2020). Implications of carbon price paths on energy security in four Baltic region countries. *Energy Strategy Reviews, 30*, 100509. https://doi.org/10.1016/j.esr.2020.100509.

Gallagher, C. L., & Holloway, T. (2020). Integrating air quality and public health benefits in US decarbonization strategies. *Frontiers in Public Health, 8*, 563358. https://doi.org/10.3389/fpubh.2020.563358.

Gilroy, J. J., Woodcock, P., Edwards, F. A., Wheeler, C., Baptiste, B. L., Uribe, C. A. M. Haugaasen, T., & Edwards, D. P. (2014) Cheap carbon and biodiversity co-benefits from forest regeneration in a hotspot of endemism. *Nature Climate Change, 4*(6), 503–507.

Griggs, D., Stafford-Smith, M., Gaffney, O., Rockström, J., Öhman, M. C., Shyamsundar, P., Steffen, W., Glaser, G., Kanie, N., & Noble, I. (2013) Sustainable development goals for people and planet. *Nature, 495*(7441), 305–307.

Harmsen, M. J., van Dorst, P., van Vuuren, D. P., van den Berg, M., Van Dingenen, R., & Klimont, Z. (2020). Co-benefits of black carbon mitigation for climate and air quality. *Climatic Change, 163*(3), 1519–1538.

Harvie, J., & Guarneri, E. (2020). Healthy people, healthy planet: Holistic thinking. In *Health of people, health of planet and our responsibility* (pp. 205–221). Cham: Springer.

He, W., Abbas, Q., Alharthi, M., Mohsin, M., Hanif, I., Vo, X. V., & Taghizadeh-Hesary, F. (2020). Integration of renewable hydrogen in light-duty vehicle: Nexus between energy security and low carbon emission resources. *International Journal of Hydrogen Energy, 45*(51), 27958–27968.

Hertwich, E., Lifset, R., Pauliuk, S., Heeren, N., Ali, S., Berrill, P., Fishman, T., Tu, Q., & Wolfram, P. (2020). Resource efficiency and climate change: Material efficiency strategies for a low-carbon future.

Howden-Chapman, P., & Chapman, R. (2012). Health co-benefits from housing-related policies. *Current Opinion in Environmental Sustainability, 4*(4), 414–419.

Howden-Chapman, P., Keall, M., Whitwell, K., & Chapman, R. (2020). The wellbeing-consumption paradox: Happiness, health, income, and carbon emissions in growing versus non-growing economies. *Science of the Total Environment, 700*, 134408.

Imai, N., Samejima, H., Langner, A., Ong, R. C., Kita, S., Titin, J., Chung, A. Y., Lagan, P., Lee, Y. F., & Kitayama, K. (2009). Co-benefits of sustainable forest management in biodiversity conservation and carbon sequestration. *PLoS ONE, 4*(12), e8267.

Intergovernmental Panel on Climate Change (IPCC). (2018). *Global research and action agenda on cities and climate change science*. Retrieved January 21, 2021, from https://www.ipcc.ch/site/assets/uploads/2019/07/Research-Agenda-Aug-10_Final_Long-version.pdf.

Intergovernmental Panel on Climate Change (IPCC). (n.d.). *Special report: Global warming of 1.5 °C—Ch 03: Impacts of 1.5 °C global warming on natural and human systems*. Retrieved December 28, 2020, Available from: https://www.ipcc.ch/sr15/chapter/chapter-3/.

Jack, D. W., & Kinney, P. L. (2010). Health co-benefits of climate mitigation in urban areas. *Current Opinion in Environmental Sustainability, 2*(3), 172–177.

Jensen, H. T., Keogh-Brown, M. R., Smith, R. D., Chalabi, Z., Dangour, A. D., Davies, M., Edwards, P., Garnett, T., Givoni, M., Griffiths, U., Hamilton, I., Jarrett, J., Roberts, I., Wilkinson, P., Woodcock, J., & Haines, A. (2013). The importance of health co-benefits in macroeconomic assessments of UK Greenhouse Gas emission reduction strategies. *Climatic Change, 121*(2), 223–237.

Jiang, P., Dong, W., Kung, Y., & Geng, Y. (2013). Analysing co-benefits of the energy conservation and carbon reduction in China's large commercial buildings. *Journal of Cleaner Production, 58*, 112–120.

Karhunmaa, K. (2016). Opening up storylines of co-benefits in voluntary carbon markets: An analysis of household energy technology projects in developing countries. *Energy Research & Social Science, 14*, 71–79.

Kim, S. E., Xie, Y., Dai, H., Fujimori, S., Hijioka, Y., Honda, Y., Hashizume, M., Masui, T., Hasegawa, T., Xu, X., Kan, Y., & Kim, H. (2020). Air quality co-benefits from climate mitigation for human health in South Korea. *Environment International, 136*, 105507.

Levy, J. I., Woo, M. K., Penn, S. L., Omary, M., Tambouret, Y., Kim, C. S., & Arunachalam, S. (2016). Carbon reductions and health co-benefits from US residential energy efficiency measures. *Environmental Research Letters, 11*(3), 034017.

Li, M., Zhang, D., Li, C. T., Mulvaney, K. M., Selin, N. E., & Karplus, V. J. (2018). Air quality co-benefits of carbon pricing in China. *Nature Climate Change, 8*(5), 398–403.

Li, N., Chen, W., Rafaj, P., Kiesewetter, G., Schöpp, W., Wang, H., Zhang, H., Krey, V., & Riahi, K. (2019). Air quality improvement co-benefits of low-carbon pathways toward well below the 2 C climate target in China. *Environmental Science & Technology, 53*(10), 5576–5584.

Li, X. (2020). Design of energy-conservation and emission-reduction plans for China's industry: Evidence from three typical industries. *Energy, 209*, 118358. https://doi.org/10.1016/j.energy.2020.118358.

Lin, B., & Yousaf Raza, M. (2020). Analysis of energy security indicators and CO_2 emissions. A case from a developing economy. *Energy, 200*, 117575. https://doi.org/10.1016/j.energy.2020.117575.

Liu, Q., & Gao, J. (2020). Public Health co-benefits of reducing greenhouse gas emissions. In *Health of people, health of planet and our responsibility* (pp. 295–307). Cham: Springer.

Lo, K., & Castán Broto, V. (2019). Co-benefits, contradictions, and multi-level governance of low-carbon experimentation: Leveraging solar energy for sustainable development in China. *Global Environmental Change, 59*, 101993.

Lu, Z., Huang, L., Liu, J., Zhou, Y., Chen, M., & Hu, J. (2019). Carbon dioxide mitigation co-benefit analysis of energy-related measures in the air pollution prevention and control action plan in the Jing-Jin-Ji region of China. *Resources, Conservation & Recycling: X, 1*, 100006. https://doi.org/10.1016/j.rcrx.2019.100006.

Luderer, G., Pehl, M., Arvesen, A., Gibon, T., Bodirsky B. L., de Boer H. S., Fricko, O., Hejazi, M., Humpenöder F., Iyer, G., Mima, S., Mouratiadou, I., Pietzcker, R. C., Popp, A., van den Berg M., van Vuuren D., & Hertwich, E. G. (2019). Environmental co-benefits and adverse side-effects of alternative power sector decarbonization strategies. *Nature Communications, 10*(1), 1–13.

Ma, D., Chen, W., Yin, X., & Wang, L. (2016). Quantifying the co-benefits of decarbonisation in China's steel sector: An integrated assessment approach. *Applied Energy, 162*, 1225–1237.

Maccagnan, A., Wren-Lewis, S., Brown, H., & Taylor, T. (2019). Wellbeing and society: Towards quantification of the co-benefits of wellbeing. *Social Indicators Research, 141*(1), 217–243.

MacKerron, G. J., Egerton, C., Gaskell, C., Parpia, A., & Mourato, S. (2009). Willingness to pay for carbon offset certification and co-benefits among (high-) flying young adults in the UK. *Energy Policy, 37*(4), 1372–1381.

Markandya, A., Armstrong, B. G., Hales, S., Chiabai, A., Criqui, P., Mima, S., Tonne, C., & Wilkinson, P. (2009). Public health benefits of strategies to reduce greenhouse-gas emissions: Low-carbon electricity generation. *The Lancet*. https://doi.org/10.1016/S0140-6736(09)61715-3.

Menikpura, S. N. M., Sang-Arun, J., & Bengtsson, M. (2013). Integrated solid waste management: An approach for enhancing climate co-benefits through resource recovery. *Journal of Cleaner Production, 58*, 34–42.

Milner, J., Davies, M., & Wilkinson, P. (2012). Urban energy, carbon management (low carbon cities) and co-benefits for human health. *Current Opinion in Environmental Sustainability, 4*(4), 398–404.

Milner, J., Hamilton, I., Woodcock, J., Williams, M., Davies, M., Wilkinson, P., & Haines, A. (2020). Health benefits of policies to reduce carbon emissions. *Bmj*, 368.

Mok, W. K., Tan, Y. X., & Chen, W. N. (2020). Technology innovations for food security in Singapore: A case study of future food systems for an increasingly natural resource-scarce world. *Trends in Food Science & Technology, 102*, 155–168.

National Oceanic and Atmospheric Administration (NOAA). (2020). *November and 2020 year to date rank 2nd hottest on record for globe.* Retrieved December 17, 2020, from https://www.noaa.gov/news/november-and-2020-year-to-date-rank-2nd-hottest-on-record-for-globe.

Nemet, G. F., Holloway, T., & Meier, P. (2010). Implications of incorporating air-quality co-benefits into climate change policymaking. *Environmental Research Letters, 5*(1), 014007.

Parry, I. W., Veung, M. C., & Heine, M. D. (2014). *How much carbon pricing is in countries' own interests? The critical role of co-benefits*. International Monetary Fund.

Paterson, M., & Stripple, J. (2010). My space: Governing individuals' carbon emissions. *Environment and Planning D: Society and Space, 28*(2), 341–362.

Peng, W., Yang, J., Lu, X., & Mauzerall, D. L. (2018). Potential co-benefits of electrification for air quality, health, and CO_2 mitigation in 2030 China. *Applied Energy, 218*, 511–519.

Phelps, J., Webb, E. L., & Adams, W. M. (2012). Biodiversity co-benefits of policies to reduce forest-carbon emissions. *Nature Climate Change, 2*(7), 497–503.

Ramaswami, A., Tong, K., Fang, A., Lal, R. M., Nagpure, A. S., Li, Y., Yu, H., Jiang, D., Russell, A. G., Shi, L., Chertow, M., Wang, Y., & Wang, S. (2017). Urban cross-sector actions for carbon mitigation with local health co-benefits in China. *Nature Climate Change, 7*(10), 736–742.

Raven, R., Sengers, F., Spaeth, P., Xie, L., Cheshmehzangi, A., & de Jong, M. (2019). Urban experimentation and institutional arrangements. *European Planning Studies, 27*(2), 258–281.

Rissel, C. E. (2009). Active travel: A climate change mitigation strategy with co-benefits for health. *New South Wales Public Health Bulletin, 20*(2), 10–13.

Rypdal, K., Rive, N., Åström, S., Karvosenoja, N., Aunan, K., Bak, J. L., Kupiainen, K., & Kukkonen, J. (2007). Nordic air quality co-benefits from European post-2012 climate policies. *Energy Policy, 35*(12), 6309–6322.

Selvakkumaran, S., & Limmeechokchai, B. (2013). Energy security and co-benefits of energy efficiency improvement in three Asian countries. *Renewable and Sustainable Energy Reviews, 20*, 491–503.

Shrestha, R. M., & Pradhan, S. (2010). Co-benefits of CO_2 emission reduction in a developing country. *Energy Policy, 38*(5), 2586–2597.

Smith, R., Guevara, O., Wenzel, L., Dudley, N., Petrone-Mendoza, V., Cadena, M., & Rhodes, A. (2019). Ensuring Co-benefits for biodiversity, climate change and sustainable development. In *Handbook of climate change and biodiversity* (pp. 151–166). Cham: Springer.

Soto-Navarro, C., Ravilious, C., Arnell, A., De Lamo, X., Harfoot, M., Hill, S. L. L., Wearn, O. R., Santoro, M., Bouvet, A., Mermoz, S., Le Toan, T., Xia, J., Liu, S., Yuan, W., Spawn, S. A., Gibbs, H. K., Ferrier, S., Harwood, T., & Kapos, V. (2020). Mapping co-benefits for carbon storage and biodiversity to inform conservation policy and action. *Philosophical Transactions of the Royal Society B, 375*(1794), 20190128.

Sovacool, B. K., Martiskainen, M., Hook, A., & Baker, L. (2020). Beyond cost and carbon: The multidimensional co-benefits of low carbon transitions in Europe. *Ecological Economics, 169*, 106529.

Sovacool, B. K., Martiskainen, M., Hook, A., & Baker, L. (2019). Decarbonization and its discontents: A critical energy justice perspective on four low-carbon transitions. *Climatic Change, 155*(4), 581–619.

Spence, C. (2014). *Global warming: Personal solutions for a healthy planet*. St. Martin's Press.

Teng, F., & Jotzo, F. (2014). Reaping the economic benefits of decarbonization for China. *China & World Economy, 22*(5), 37–54.

Thompson, T. M., Rausch, S., Saari, R. K., & Selin, N. E. (2014). A systems approach to evaluating the air quality co-benefits of US carbon policies. *Nature Climate Change, 4*(10), 917–923.

Thompson, T. M., Rausch, S., Saari, R. K., & Selin, N. E. (2016). Air quality co-benefits of subnational carbon policies. *Journal of the Air & Waste Management Association, 66*(10), 988–1002.

Tonmoy, F. N., Cooke, S. M., Armstrong, F., & Rissik, D. (2020). From science to policy: Development of a climate change adaptation plan for the health and wellbeing sector in Queensland, Australia. *Environmental Science & Policy, 108*, 1–13.

Trotta, G. (2020). Assessing energy efficiency improvements and related energy security and climate benefits in Finland: An ex post multi-sectoral decomposition analysis. *Energy Economics, 86*, 104640. https://doi.org/10.1016/j.eneco.2019.104640.

United Nations Environment Programme (UNEP). (2020). *Emissions Gap Report 2020.* Retrieved January 21, 2021 from https://www.unenvironment.org/emissions-gap-report-2020.

West, J. J., Smith, S. J., Silva, R. A., Naik, V., Zhang, Y., Adelman, Z., Fry, M. M., Anenberg, S., Horowitz, L. W., & Lamarque, J. F. (2013). Co-benefits of mitigating global greenhouse gas emissions for future air quality and human health. *Nature Climate Change, 3*(10), 885–889.

Wittman, H. K., & Caron, C. (2009). Carbon offsets and inequality: Social costs and co-benefits in Guatemala and Sri Lanka. *Society and Natural Resources, 22*(8), 710–726.

Zhang, S., Worrell, E., & Crijns-Graus, W. (2015). Evaluating co-benefits of energy efficiency and air pollution abatement in China's cement industry. *Applied Energy, 147*, 192–213.

Zhang, S., Worrell, E., Crijns-Graus, W., Wagner, F., & Cofala, J. (2014). Co-benefits of energy efficiency improvement and air pollution abatement in the Chinese iron and steel industry. *Energy, 78*, 333–345.

Zhou, Y., Ma, M., Kong, F., Wang, K., & Bi, J. (2018). Capturing the co-benefits of energy efficiency in China—A perspective from the water-energy nexus. *Resources, Conservation and Recycling, 132*, 93–101.

Chapter 4
Low Carbon Transitions: A Global Overview

4.1 An Insight on Low Carbon Transitions: China and the Globe

Driven by the global response to climate change, the world is undergoing changes in its economic and social development patterns. As discussed in the previous chapter, the impacts of climate change are quite severe. It remains a question whether the consequences are irreversible or not? Or are we too late to implement concrete action plans, which are realistic to the conditions of today and tomorrow? Could we move towards a feasible low carbon transition that could transform the way our cities, communities, planning, and policies are now? To answer these questions, we delve into further discussions on low carbon transitions, looking into the diversity of low carbon directions, tools, and opportunities (Crawford & French, 2008; Cheshmehzangi, 2020a, 2020b; Roberts et al., 2018). In the age of urban entrepreneurship (Xie et al., 2020), we have to consider pathways towards sustainable development and those that could lead us through feasible transitions and new modes of development (Cohen & Muñoz, 2015; Minniti, 2013; Rodrigues & Franco, 2018; Skokic & Morrison, 2011). We have to understand the role of actors, the availability of factors, and the willingness to make progress through experimentation and innovation (Cheshmehzangi et al., 2018). In this chapter, we aim to address these from the perspective of 'low carbon transitions'.

Low carbon economy, low carbon development, low carbon technology, and other concepts began to rise in various countries. This has been ongoing for the last two or three decades. The progress may appear slow, but it covers a wide range of sectors, disciplines, policies, and applications. For example, the development of low carbon energy technology improves energy efficiency, optimises the energy structure, and changes economic development mode. It also establishes a low carbon economic development model and a low carbon social consumption model. The world's low carbon transition has triggered new economic, trade, and technological competition.

A. Cheshmehzangi and H. Chen, *China's Sustainability Transitions*,
https://doi.org/10.1007/978-981-16-2621-0_4

Thus, the approach towards low carbon goals simply makes sense. The new directions, combined with the continuing urbanisation trends, imply integrated sustainable strategies, hybrid models of development, and more comprehensive methods in achieving sustainability ideals.

Many countries have formulated national strategies for adaptation to climate change according to their own conditions and achieve a sustainable economy, society, and environment. Many global examples incorporate sustainable development to combat climate change and its effects on our living and natural habitats (Cheshmehzangi & Dawodu, 2019a). The low carbon economy transition in developed countries is mainly to reduce the current carbon emission level through technological innovation and economic and social transformation on the premise of maintaining the current high level of economic development and social consumption (Cheshmehzangi et al., 2021; Deng & Cheshmehzangi, 2018; Garnaut, 2010; He, 2016a; Mulugetta & Urban, 2010; Nyambuu & Semmler, 2020; Urban, 2014). Similar trends are also seen in China's early stages of low carbon economic transition (Yan et al., 2019; Zhang, 2010). China's progress will be studied further throughout the book. In doing so, we also compare commonalities and differences between China and other countries, especially those that have adapted low carbon strategies or are already in the process of low carbon transitions.

As the global response to climate change becomes increasingly urgent, China is confronted with challenges and opportunities brought about by the global response to climate change and the world's transition to a low carbon economy (Barbi et al., 2016; Chen, 2012; Cheshmehzangi, 2016; Gilley, 2012; Heggelund, 2007; Lewis, 2009; Song & Woo, 2008). Low carbon development is also identified as an important strategy to promote energy technology innovation, transform economic development model, and protect global climate relations (Beermann, 2014; Byrne et al., 2011; Cheshmehzangi et al., 2021; Meckling & Allan, 2020; Shen & Sun, 2016; Wang & Chang, 2014). Therefore, we can verify that low carbon development is an inevitable trend. In this pathway, China and many other countries in the world promote cooperation to achieve the win–win situation of tackling global climate change and attain domestic sustainable development (He, 2016b). In this chapter, we explore some of these pathways as part of the process of low carbon transitions. We first look into various available strategies, specifically from adaptation and mitigation viewpoints, in combatting climate change issues. Afterward, we delve into low carbon transitions and explore some of the existing initiatives worldwide. This is done followed by brief analyses of low carbon transitions globally and in China. Both contexts are explored through brief analyses before explaining China's commitment to the carbon neutrality plan and its goals. This discussion view is then carried out for the rest of the chapter, through which we relate the topic to the sustainable development goals (SDGs), low carbon targets, and China's target of 2060.

4.2 Adaptation and Mitigation Strategies to Climate Change: Towards Carbon Neutrality

At first, we aim to continue our discussions on climate change impacts and how low carbon transitions are considered inevitable should we wish to reverse some of the current and recent trends. Before the 2 °C global temperature target was proposed, some international authorities, which were led by the European Union, had already carried out certain scientific research and exploration on climate change (Cheshmehzangi & Dawodu, 2019b; Den Elzen et al., 2006; Gao et al., 2017; Kriegler et al., 2013; Meinshausen et al., 2009; New et al., 2011; Raftery et al., 2017; Randalls, 2010; Tol, 2007; Vautard et al., 2014). In 1996, the Council of the European Union proposed the goal of keeping the global average surface temperature within 2 °C of pre-industrial levels (EU Council of Ministers, 1996). The 2007 Fourth Assessment Report (AR4) of the Intergovernmental Panel on Climate Change (IPCC), which laid the groundwork for the post-Kyoto agreement, provided scientific evidence for the decision to limit temperature rise to 2 °C (Füssel, 2009; Pachauri & Reisinger, 2007; Solomon et al., 2007). In 2009, the global 2 °C temperature target was written into the Copenhagen Accord after the consensus of major developed and developing countries (Houser, 2010; Rajamani, 2010; Randalls, 2010; UNFCCC, 2009). However, the Parties to the United Nations Framework Convention on Climate Change (UNFCCC) have not unanimously recognised the Copenhagen Accord since the Cancun Climate Change Conference in 2010 (UNFCCC, 2010). Then the IPCC Fifth Assessment Report (AR5) in 2013–2014 further confirmed and supported the conclusions of AR4 on the impact of human activities from multiple perspectives (IPCC,) (Table 4.1).

Table 4.1 Different strategies to climate change

Adaptation strategies	Mitigation strategies
UK adapting to climate change, 2008 Scotland adapting our way: managing Scotland's climate risk, 2008 Finland national strategy for adaptation to climate change, 2005 Germany adaptation strategy for climate change, 2008 France Adaptation strategy for climate change, 2005 Australia national climate change adaptation framework, 2007 European Commission adaptation to climate change in Europe—EU action options, 2007 European Commission White Paper on adaptation to climate change: a framework for action for Europe, 2009	United Nations framework convention on climate change (UNFCCC), 1994 Nairobi programme of work, 2005 Bali action plan 2007 The United States Kyoto Protocol, 2001

Note Other strategies may exists from other countries and here we mainly provide a selection of different strategies

At the same time, it provided the cumulative emissions under the 2 °C temperature target. It also gave the risk assessment of climate change in different areas and regions under different future temperature rise scenarios, as well as emission scenarios for controlling the 2 °C temperature rise. Since some countries believe that the 2 °C target is insufficient to protect them from the threat of rising sea levels and global warming, the global average temperature rise of 1.5 °C was argued to be the target (Cheshmehzangi, 2020c; Huntingford & Mercado, 2016; Knopf et al., 2012; Magnan et al., 2016; Michaelowa et al., 2018; Sanderson et al., 2016; Schellnhuber et al., 2016; Tanaka & O'Neill, 2018; Taylor et al., 2018). The Paris Agreement in 2015 formally set the goal of limiting the global average temperature rise to less than 2 °C from the pre-industrial level by the end of the century. It made efforts to keep it within 1.5 °C (United Nations, 2015). The Paris Agreement was the first international treaty to give legal force to the global 2 °C target.

Keeping the global average temperature rise below 1.5 °C depends on the ability of human society to achieve transformation in energy, land, cities, infrastructure, and industry (Butters et al., 2020; Cheshmehzangi and Dawodu, 2019a; Jones, 2018; Jewel & Cherp, 2020; Leng, 2018; Park, 2018; Tollefson, 2018; Rogelj et al., 2015; Rogelj et al., 2018). In sum, limiting to 1.5 °C would require a 45% drop in anthropogenic carbon dioxide emissions from 2010 levels by 2030 (or 40–60% interquartile range) and Net Zero Emissions by around 2050 (or 2035–2055 interquartile range) (IPCC, 2019). Therefore, compared with a temperature rise of 2 °C, the 1.5 °C target requires humankind to make more efforts than achieving the target of 2 °C set in the Paris Agreement. In this regard, low carbon development and transitions become more effective in the coming decades.

At the UN Climate Action Summit in September 2019, 66 countries and more than 100 local governments announced the goal of achieving net zero carbon emissions by the middle of this century (As shown in Table 4.2) (UNFCCC, 2019). The Energy & Climate Intelligence Unit's net zero emissions tracker charts the progress of countries. Two have achieved it (Suriname and Bhutan). Six have it in law (Sweden, UK, France, Denmark, etc.). The EU, Canada, Chile, and Fiji are currently in the process of legislating it. Besides, 12 countries, including China and some European Union countries, have issued policy statements (Fig. 4.1). For example, In December 2019, the German Bundestag passed the Federal Climate Protection Law (see Climate Change Laws of the World, n.d.), which set the mid- long term emission reduction target of reducing greenhouse gas emissions by 55% by 2030 compared with 1990 and achieving net zero emissions by 2050 (Hu & Qiu, 2019; Jäger-Waldau et al., 2020; Tanneberger et al., 2021; Tsiropoulos et al., 2020). The law decomposes the German federal emission reduction targets in energy, industry, construction, transportation, agriculture, and forestry. It stipulates the legal mechanism for departments and their emission reduction measures, adjustment of emission reduction targets, and regular assessment of emission reduction effects. Other examples from the EU have similar targets (Karlsson et al., 2020; Lund & Matheisen, 2009; Millot et al., 2020; Rehfeldt et al., 2020; Steininger et al., 2020; Zaklan et al., 2021), showing the scale of low carbon targets and strategies.

Table 4.2 Carbon neutral targets of countries or regions (Data is extracted from https://climatenewsnetwork.net/ and https://www.iea.org/countries)

Country	Target year	Nature of commitment	Content
Austria	2040	Policy statement	Committed to climate neutrality by 2040 and 100% clean electricity by 2030 based on the 2017 climate policies
Bhutan	Carbon Negative	Achieved	Bhutan has a population of less than one million and is surrounded by forests and hydropower resources. It is claimed to be the first carbon negative country of the twenty-first century
Canada	2050	Proposed legislation	Committed to a net zero emission target. It has also developed a legally binding five-year carbon budget
Chile	2050	Proposed legislation	In April 2020, an enhanced medium-term commitment was submitted to the United Nations to close eight of the 28 coal-fired power plants by 2024 and phase out coal-fired power by 2040
China	2060	Policy statement	On September 22, 2020, China announced to the United Nations General Assembly that it would strive to achieve carbon neutrality in 2060 and reach the peak emission by 2030
Costa Rica	2050	Policy statement	Submitted to the United Nations in December, it plans to have zero net emissions by 2050

(continued)

Table 4.2 (continued)

Country	Target year	Nature of commitment	Content
Denmark	2050	Legal provisions	A plan to build a "climate neutral society" by 2050 was formulated in 2018, banning the sale of new gasoline and diesel vehicles from 2030, and supporting electric vehicles
European Union (EU)	2050	Proposed legislation	The "green agreement" announced in December 2019 proposes the EU's goal of net-zero emissions by 2050, and the long-term strategy was submitted to the United Nations in March 2020
Fiji	2050	Proposed legislation	A plan was submitted to the United Nations in 2018 with the goal of achieving zero net carbon emissions in all sectors of the economy
Finland	2035	Policy statement	Strengthen climate law in June 2019. It will require restrictions on industrial logging and a phasing out of peat burning for power generation
France	2050	Legal provisions	On June 27, 2019, the French National Assembly voted to bring the net zero target into law. The High Commission on climate proposed that France must triple its emission reduction rate in order to achieve the goal of carbon neutrality

(continued)

Table 4.2 (continued)

Country	Target year	Nature of commitment	Content
Germany	2050	Policy statement	Germany's first major climate law came into effect in December 2019, proposing to neutralise greenhouse gases by 2050
Hungary	2050	Legal provisions	Hungary committed to climate neutrality by 2050 in the climate law passed in June 2020
Iceland	2040	Policy statement	Iceland has already obtained almost carbon free electricity and heating from geothermal and hydroelectric power generation, and the strategy announced in 2018 focuses on phasing out fossil fuels for transportation, planting trees and restoring wetlands
Ireland	2050	Policy Statement	In the joint agreement in June 2020, it is proposed to legally set the goal of net-zero emission by 2050 and reduce emissions by 7% annually in the next 10 years
Japan	The second half of this century	Policy statement	In June 2019, the Japanese government approved a climate strategy focusing on carbon capture, utilisation and storage, as well as the development of hydrogen as a source of clean fuel. Coal is expected to still supply a quarter of the country's electricity by 2030

(continued)

Table 4.2 (continued)

Country	Target year	Nature of commitment	Content
Marshall Islands	2050	Policy statement	The latest report submitted to the United Nations in September 2018 sets out the desire to achieve net zero emissions by 2050
New Zealand	2050	Legal provisions	A law passed in November 2019 sets a net-zero target for all greenhouse gases except bio-methane
Norway	2050	Policy Statement	To be carbon neutral by 2030 through international offsets and to be carbon neutral domestically by 2050
Portugal	2050	Policy statement	Portugal is a world leader in integrating wind and solar photovoltaic power, and has met its goal of generating 80% of its electricity from renewable sources by 2030 and a carbon–neutral economy by 2050
Singapore	The second half of this century	Submission to the United Nations	Singapore has set strong targets including increasing the country's energy efficiency by 36% by 2030, in compared to the 2005 levels. Singapore introduced energy efficiency standards and labels for lamps in 2015. The government also plans to increase solar photovoltaic capacity, reduce greenhouse gas emissions and peak national emissions by 2030

(continued)

Table 4.2 (continued)

Country	Target year	Nature of commitment	Content
Slovenia	2050	Policy statement	Slovenia's Low Carbon Strategy draft, presented in September 2011, aims to achieve a high quality of life, space and natural environment by 2050
Slovakia	2050	Submission to the United Nations	The key objectives of Slovakia's energy policy agenda are to improve efficiency in the areas of capacity and end-use, reduce energy intensity, reduce dependence on energy imports, expand the use of nuclear energy, increase the share of renewable energy in the thermal and electricity sectors, and support alternative fuel use for transport
South Africa	2050	Policy statement	Coal is the backbone of South Africa's energy system, supplying about 70% of its installed generating capacity. However, the 2019 Integrated Resource Plan sets the goal of long-term diversification of the power mix by 2030
South Korea	2050	Proposed legislation	In 2012, South Korea announced an emissions trading scheme, the first of its kind in Asia. In 2020, South Korea outlined plans to decarbonise its economy by 2050 and end coal financing, which is the first such commitment in East Asia

(continued)

Table 4.2 (continued)

Country	Target year	Nature of commitment	Content
Spain	2050	Proposed legislation	Spain has built up a strong electricity system, with a high proportion of wind and solar photovoltaic power generation. A draft climate framework bill was submitted in May 2020
Sweden	2045	Legal provisions	The targets which are set in the Energy Agreement and Climate Framework aim to become a net-zero carbon economy by 2045
Switzerland	2050	Policy statement	Switzerland's carbon-free electricity sector is dominated by nuclear and hydroelectric power. The Swiss Federal Council announced on August 28, 2019 that it would achieve net-zero carbon emissions by 2050
UK	2050	Legal provisions	The UK has set a target to reduce emissions, a net zero emission target by 2050 and a carbon budget. Wind and solar power are expected to reach more than 50% by 2030
Uruguay	2030	Submission to the United Nations	Reduce beef farming/agriculture, waste and energy emissions to become a net-zero nation by 2030

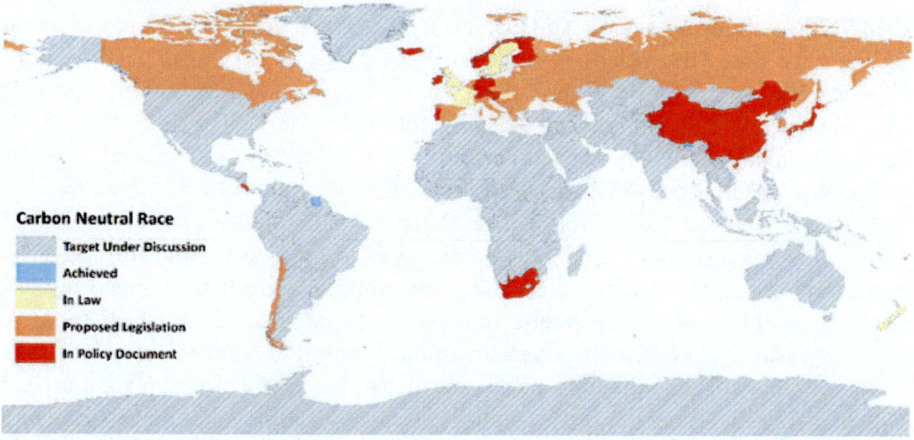

Fig. 4.1 Carbon neutral race of different countries and regions (Drawn by the Authors, data is extracted from https://eciu.net/netzerotracker)

4.3 Low Carbon Transition: Exploring Some of the Existing Initiatives

The measures of low carbon transition in various countries mainly focus on putting forward transition planning, optimising energy systems, and developing a circular economy. For instance, the EU issued a Green New Deal at the end of 2019 (Chohan, 2019; Mastini et al., 2021; Maya-Drysdale et al., 2020; Pettifor, 2020), which included a roadmap for policies and measures in seven areas, including energy, industry, and transport. Global coal consumption has declined significantly since 2013 (Guan et al., 2018; Liu et al., 2020; Tang et al., 2018; Wang et al., 2020), with the share of coal in primary energy consumption falling from 37% in 1965 to 27% in 2019. Nearly 180 countries have set goals or policies to promote the development of renewable energy. Scientific and technological innovation is an essential strategy for developed countries/regions such as the United States and in the European countries. These innovations are mainly considered to tackle climate change. For example, the EU proposed supporting research and innovation through its $94.1 billion research and development program, Horizon Europe (2021–2027). The EU issued a new circular economy action plan in March 2020, which aims to implement the circular economy concept throughout the whole life cycle of products and reduce resource consumption and carbon footprint. The US has also proposed promoting cutting-edge technological innovation such as carbon capture, next-generation nuclear energy, and electric vehicles.

To be specific, some typical low carbon cities in the world mainly realise the low carbon transformation from the aspects of Power, Transportation, Industry, Buildings, Agriculture, etc. Copenhagen, the capital of Denmark, is a model for developing a low carbon economy (Amer et al., 2019; Can et al., 2011; Lewis, 2012; Messner et al., 2013). Denmark has increased its renewable energy capacity from 5% to more than 70% in about 25 years (XinHuaNet, 2020). The Copenhagen government set up the City Bike (Bycyklen) system in 1995 (DMC, 2021). From 1990 to 2007, Copenhagen promoted low carbon concepts in energy production, energy consumption, and energy-efficient buildings (Hald-Mortensen, 2009). The CPH climate plan 2025, which includes energy consumption, energy production, commuting, and aims to become the first carbon–neutral city by 2025, was launched in Copenhagen in 2012 (Damsø et al., 2017; Horswill & Nielsen, 2016; Madsen & Hansen, 2019). In its climate action plan, Copenhagen proposes to build 100 wind turbines; Heat consumption and commercial electricity consumption were both reduced by 20%; Cycling, walking, or taking public transport accounted for 75% of all trips; All organic waste to achieve biomass gasification; Erecting 60,000 m^2 of solar panels; 100% of Copenhagen's heating needs are met by renewable energy sources (C40 Cities, 2013). Each person in Copenhagen now owns a bicycle, the city has more than 500 km of cycle lanes, and nearly a third of people cycle to work. The development of energy-efficient buildings, the development of new energy sources, the increase of waste recycling, and forest areas' expansion are also energy conservation and emission reduction measures.

In Amsterdam, the Netherlands, bicycles are the most common form of transportation. Amsterdam's city government launched the Smart City project in 2009 (Raven et al., 2019). The main goal is to reduce carbon dioxide emissions and save energy. They are encouraging recycling, installing solar panels, and providing more electric car and bus services. Most residents of Amsterdam generate power/energy from solar panels and small wind turbines. Most homes have also installed energy-saving systems to reduce usage and save electricity (Ou, 2017). In another example, Stockholm, Sweden, is dedicated to green living and cleaning. The government focuses on increasing biogas production to reduce reliance on fossil fuels, with plans to make fossil fuels free for urban use by 2050 (Gustavsson, 2011). The city has many fuel cell bus services that provide clean urban transportation. There are also hundreds of bike stations around the city. Vancouver, Canada, relies on renewable energy sources to minimise carbon dioxide emissions, generating 93% of the city's electricity needs from renewable hydropower (Bać, 2014). Reykjavik, the capital of Iceland, has the largest geothermal heating system in the world. There are 30 active volcanoes in the country. The high availability of geothermal energy in the city also meets most residents' hot water needs, and the use of these renewable energy sources also protects the city from excessive carbon dioxide emissions (Kennedy, 2014).

The London Array is the largest offshore wind farm globally, meeting the needs of 750,000 homes (DULAS, 2021). Aggressive promoting renewable energy and new green projects will also make the city carbon–neutral over the next decade. San Francisco, USA, uses a high recycling level, with waste used directly for recycling rather than garbage. San Francisco actively encourages the use of renewable energy and has many community welfare programs. Initiatives such as biking, ride-sharing, e-mobility, and city-ride sharing can save money and reduce greenhouse gas emissions (CEEX, 2018). They have become ubiquitous initiatives now adopted by many cities and countries with a low carbon agenda. These examples are to highlight some global initiatives, perhaps common in many contexts, before we delve into the global transition pathways in the following section.

4.4 A Brief Analysis of Global Examples

In this section, we highlight some changes in GHG emissions of major countries of the world. In doing so, the top ten greenhouse gas emissions countries in the world are selected for analysis. The total national emissions in the nearly past 30 years are shown in Fig. 4.2. As shown in these records, China is the country with the highest GHG emissions globally, with the most significant increase of approximately three times in compared to 1990. The other higher increases are followed by Iran and India, at 2.4 times and 1.8 times, respectively. On the contrary, Russia, Germany, the European Union (inclusive of 27 countries), and Brazil had a decrease in their GHG emissions by 30.88%, 28.41%, 24.73%, and 6.92%, respectively compared with 1990. From the findings, we noted it was evident that the top ten greenhouse gas emissions countries have remained stable in the past five years. Some examples include China at 11,580 MtCO_2e and United States at 6000 MtCO_2e.

From the perspective of per capita greenhouse gas (GHG) emissions, most of the top ten emitters are higher than the world average (about 6.5 tCO_2e per capita). Based on the records, only India has not exceeded significantly from 1990 to 2017 (Fig. 4.3). Among them, Canada, the United States, and Russia have the highest

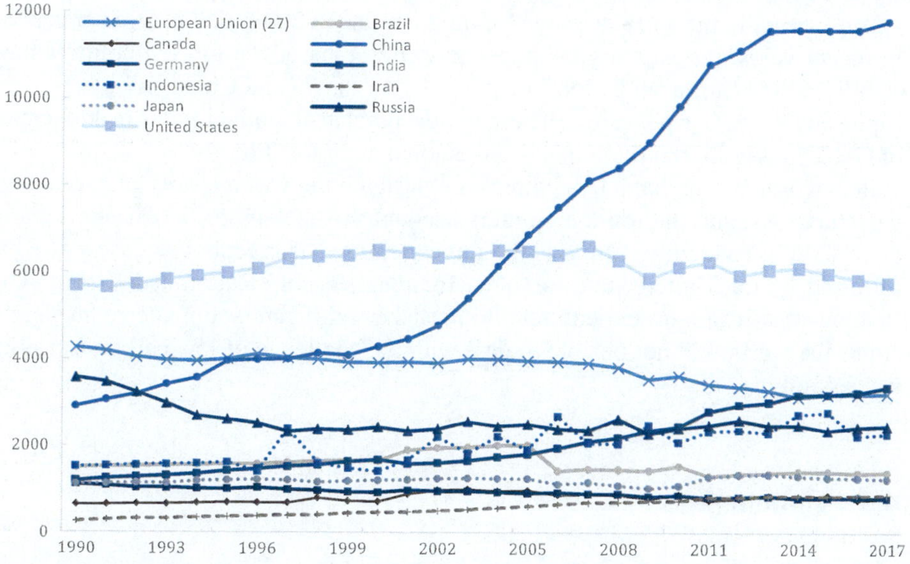

Fig. 4.2 The total GHG emissions of the world's major countries and economies from 1990–2017(WRI) (Unit: MtCO2e. Data Resource from: www.wri.org)

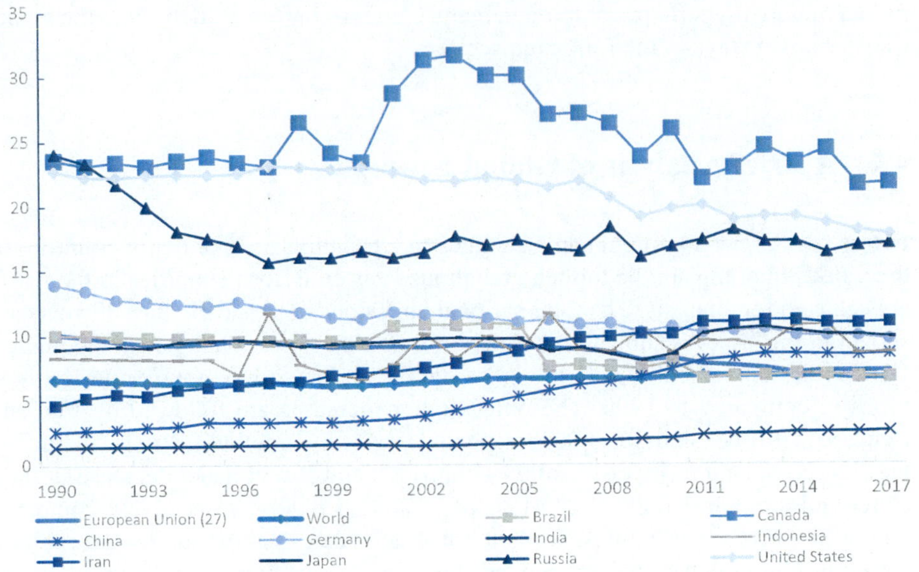

Fig. 4.3 Per capita GHG emissions of the world's major countries and economies from 1990–2017 (WRI) (Unit: tCO_2e per capita. Data Resource from: www.wri.org)

per capita greenhouse gas emissions, with 21.7 tCO2e per capita, 17.75 tCO_2e per capita, and 17.03 tCO_2e per capita, respectively. For instance, China's per capita emissions have increased year by year until 2013, with an average annual increase of 8.3%, exceeding the world average for the first time in 2009 (6.75tCO_2e per capita). In recent years, per capita greenhouse gas emissions of the top ten emitters have stabilised, like European Union (including its 27 states) and China at 7.2 tCO_2e per capita and 8.5 tCO_2e per capita. These trends indicate potential threats in an increase of GHG emissions and the impacts on regional and global levels.

It is worth noting that a few countries contribute the vast majority of greenhouse gas emissions, and the top ten emitters account for 62% in 2017 (Cheshmehzangi et al., 2021). Evidently, China is the largest emitter of greenhouse gases, at 24%, followed by the United States at 11%, India at 7%, and Russia at 5% (Fig. 4.4). It shows that if the top ten emitters do not take significant actions to reduce emissions, the world will not be able to deal with the challenge of low carbon transition successfully.

4.5 Conclusions

An inevitable wave of the global green and low carbon transformation is emerging. Under the framework of the Paris Agreement, countries are required not only to submit near-medium term "nationally determined contributions" climate targets but

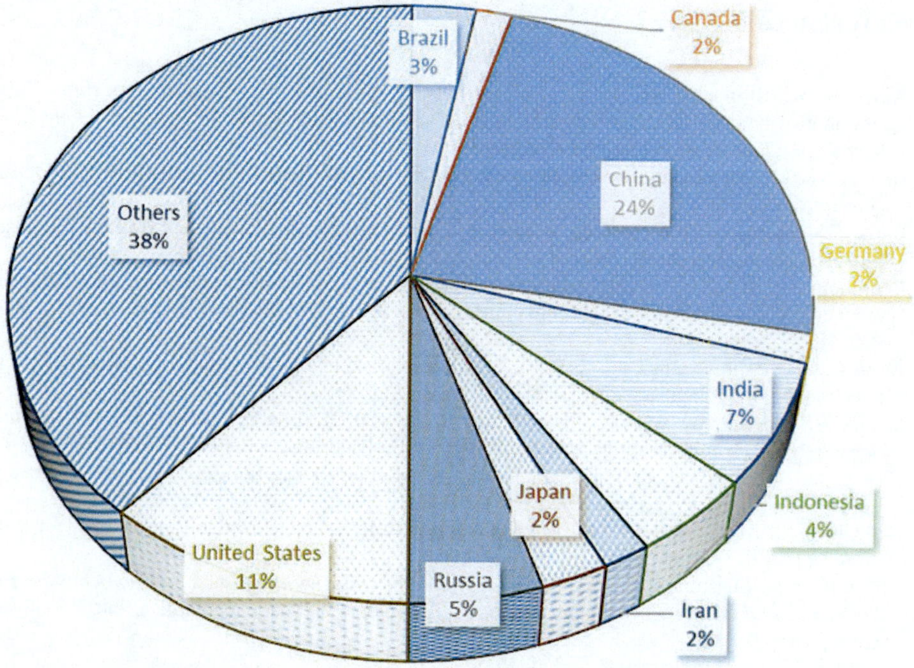

Fig. 4.4 Proportion of GHG emissions of the world's major countries and economies in 2017 (WRI) (Data Resource from: www.wri.org)

also to formulate long-term low emission strategies up to the middle of the century. At present, many countries are committed to exploring a feasible path of net-zero emissions in line with national conditions. In September 2020, China announced that it would be carbon neutral by around 2060, followed by Japan and South Korea, which announced that they would be carbon neutral by 2050. As of October 2020, participants in the net-zero goal cover 826 cities, 103 regions, and 1565 companies worldwide, all of which represent 880 million residents, 24.9 million employees, and 10 billion tons of greenhouse gas emissions (New Climate Institute, 2020). As countries or regions begin to implement their plans, they will promote investment in renewable energy, smart buildings, and green transport to mitigate climate change. In particular, some major carbon emitters, which account for the bulk of global emissions, need to step up their level of action. They also need to make greater efforts to build resilience in vulnerable countries that emit less carbon but are most vulnerable to climate change.

References

Amer, S. B., Bramstoft, R., Balyk, O., & Nielsen, P. S. (2019). Modelling the future low-carbon energy systems-case study of Greater Copenhagen, Denmark. *International Journal of Sustainable Energy Planning and Management, 24.*

Bać, A. (2014). Sustainable development of the city on the example of Vancouver. *Ecological Questions, 19*, 83.

Barbi, F., Ferreira, L. D. C., & Guo, S. (2016). Climate change challenges and China's response: Mitigation and governance. *Journal of Chinese Governance, 1*(2), 324–339.

Beermann, J. (2014). Urban partnerships in low-carbon development: Opportunities and challenges of an emerging trend in global climate politics. *urbe. Revista Brasileira de Gestão Urbana, 6*(2), 170–183.

Butters, C., Cheshmehzangi, A., & Sassi, P. (2020). Cities, energy and climate: Seven reasons to question the dense high-rise city. *Journal of Green Building, 15*(3), 197–214.

Byrne, R., Smith, A., Watson, J., & Ockwell, D. (2011). Energy pathways in low-carbon development: From technology transfer to socio-technical transformation.

Can, H., & Hongbing, D. (2011). The model of developing low-carbon tourism in the context of leisure economy. *Energy Procedia, 5*, 1974–1978.

C40 Cities. (2013). *Copenhagen: CPH climate plan 2025.* Webpage available at: https://www.c40.org/profiles/2013-copenhagen

CEEX (China Emissions Exchange). (2018). *The eight cities with the lowest carbon emissions in the world, and what environmental measures they adopt*. Webpage available at: https://mp.weixin.qq.com/s/SN8UGc1yygR43-nQGemckg

Chen, G. (2012). *China's climate policy*. Routledge.

Cheshmehzangi, A. (2016). China's New-Type Urbanisation Plan (NUP) and the foreseeing challenges for decarbonization of cities: A review. *Energy Procedia, 104*, 146–152.

Cheshmehzangi, A. (2020a). Low carbon transition at the township level: Feasibility study of environmental pollutants and sustainable energy planning. *International Journal of Sustainable Energy*, 1–27.

Cheshmehzangi, A. (2020). Towards urban energy management: A brief review of China's recent transition in urban energy demand. *Low Carbon Economy, 11*(1), 1–24.

Cheshmehzangi, A. (2020). The analysis of global warming patterns from 1970s to 2010s. *Atmospheric and Climate Sciences, 10*(3), 392–404.

Cheshmehzangi, A., & Dawodu, A. (2019). *Sustainable urban development in the age of climate change—People: Cure or curse.* Palgrave Macmillan.

Cheshmehzangi, A., & Dawodu, A. (2019b). The review of sustainable development goals (sdgs): people, perspective and planning. In *Sustainable urban development in the age of climate change—People: Cure or curse* (pp. 133–156). Palgrave Macmillan.

Cheshmehzangi, A., Dawodu, A., & Sharifi, A. (2021). *Sustainable urbanism in China.* Routledge.

Cheshmehzangi, A., Xie, L., & Tan-Mullins, M. (2018). The role of international actors in low-carbon transitions of Shenzhen's international low carbon city in China. *Cities, 74*, 64–74.

Chohan, U. W. (2019). A green new deal: Discursive review and appraisal. *Notes on the 21st Century (CBRI*).

Climate Change Laws of the World. (n.d.). *Law to introduce a federal climate protection law and to change further regulations* ("Bundesklimaschutzgesetz" or "KSG"). Available in LSE archive at: https://climate-laws.org/geographies/germany/laws/law-to-introduce-a-federal-climate-protection-law-and-to-change-further-regulations-bundesklimaschutzgesetz-or-ksg

Cohen, B., & Muñoz, P. (2015). Toward a theory of purpose-driven urban entrepreneurship. *Organization & Environment, 28*(3), 264–285.

Crawford, J., & French, W. (2008). A low-carbon future: Spatial planning's role in enhancing technological innovation in the built environment. *Energy Policy, 36*(12), 4575–4579.

Damsø, T., Kjær, T., & Christensen, T. B. (2017). Implementation of local climate action plans: Copenhagen-towards a carbon-neutral capital. *Journal of Cleaner Production, 167*, 406–415.

Denmark Media Centre (DMC). (2021). *Bycyklen.* Webpage available at: https://www.visitdenmark.com/denmark/explore/bycyklen-gdk495345

Den Elzen, M., & Meinshausen, M. (2006). Meeting the EU 2 C climate target: Global and regional emission implications. *Climate Policy, 6*(5), 545–564.

Deng, W., & Cheshmehzangi, A. (2018). *Eco-development in China: Cities, Communities and buildings.* Palgrave Macmillan.

DULAS. (2021). *London array wind farm wind monitoring instrumentation.* Webpage available at: https://dulas.org.uk/london-array-wind-farm-wind-monitoring-instrumentation/

EU Council of Ministers. (1996). Community strategy on climate change—Council conclusions. In *Decision of the Council of the European Union (EU) conference in 1996*, concluded on 25 June 1996; held in Luxembourg, Luxembourg.

Füssel, H. M. (2009). An updated assessment of the risks from climate change based on research published since the IPCC fourth assessment report. *Climatic Change, 97*(3), 469–482.

Gao, Y., Gao, X., & Zhang, X. (2017). The 2 C global temperature target and the evolution of the long-term goal of addressing climate change—From the United Nations framework convention on climate change to the Paris agreement. *Engineering, 3*(2), 272–278.

Garnaut, R. (2010). Policy framework for transition to a low-carbon world economy. *Asian Economic Policy Review, 5*(1), 19–33.

Gilley, B. (2012). Authoritarian environmentalism and China's response to climate change. *Environmental Politics, 21*(2), 287–307.

Guan, D., Meng, J., Reiner, D. M., Zhang, N., et al. (2018). Structural decline in China's CO_2 emissions through transitions in industry and energy systems. *Nature Geoscience, 11*(8), 551–555.

Gustavsson, M., Särnholm, E., Stigson, P., & Zetterberg, L. (2011) *Energy scenario for Sweden 2050. Swedish Environmental Research Institute.* Available from: https://www.researchgate.net/publication/285766198_Energy_Scenario_for_Sweden_2050_Based_on_Renewable_Energy_Technologies_and_Sources

Hald-Mortensen, C. (2009). *Transatlantic climate policy: Towards a Copenhagen protocol in 2009.* Available from: https://www.bioin.or.kr/InnoDS/data/upload/policy/1233213846375.pdf

He, J. K. (2016). Global low-carbon transition and China's response strategies. *Advances in Climate Change Research, 7*(4), 204–212.

He, J. (2016b). New governance mechanism for global climate change and China's economic low-carbon transformation. *Wuhan University Journal (Philosophy & Social Sciences), 04.*

Heggelund, G. (2007). China's climate change policy: Domestic and international developments. *Asian Perspective*, 155–191.

Horswill, D. G., & Nielsen, T. H. (2016). Can CLT construction help Copenhagen become world's first carbon neutral city? *Structures and Architecture: Beyond their Limits, 153.*

Houser, T. (2010). *Copenhagen, the accord, and the way forward (No. PB10–5).* Washington, DC: Peterson Institute for International Economics.

Hu, M., & Qiu, Y. (2019). A comparison of building energy codes and policies in the USA, Germany, and China: Progress toward the net-zero building goal in three countries. *Clean Technologies and Environmental Policy, 21*(2), 291–305.

Huntingford, C., & Mercado, L. M. (2016). High chance that current atmospheric greenhouse concentrations commit to warmings greater than 1.5 °C over land. *Scientific Reports, 6*(1), 1–7.

Intergovernmental Panel on Climate Change (IPCC). (2019). *Global warming of 1.5 °C, headline statements from the summary for policymakers* (2 pages). Available from: https://www.ipcc.ch/site/assets/uploads/sites/2/2019/06/SR15_Headline-statements.pdf

Jäger-Waldau, A., Kougias, I., Taylor, N., & Thiel, C. (2020). How photovoltaics can contribute to GHG emission reductions of 55% in the EU by 2030. *Renewable and Sustainable Energy Reviews, 126*, 109836.

Jewell, J., & Cherp, A. (2020). On the political feasibility of climate change mitigation pathways: Is it too late to keep warming below 1.5 °C? *Wiley Interdisciplinary Reviews: Climate Change, 11*(1), e621.

Jones, A., Haywood, J. M., & Jones, C. D. (2018). Can reducing black carbon and methane below RCP2. 6 levels keep global warming below 1.5 °C? *Atmospheric Science Letters, 19*(6), e821.

Karlsson, I., Rootzén, J., & Johnsson, F. (2020). Reaching net-zero carbon emissions in construction supply chains–Analysis of a Swedish road construction project. *Renewable and Sustainable Energy Reviews, 120*, 109651.

Kennedy, C. A., Ibrahim, N., & Hoornweg, D. (2014). Low-carbon infrastructure strategies for cities. *Nature Climate Change, 4*(5), 343–346.

Knopf, B., Kowarsch, M., Flachsland, C., & Edenhofer, O. (2012). The 2 C target reconsidered. In *Climate change, justice and sustainability* (pp. 121–137). Springer.

Kriegler, E., Tavoni, M., Aboumahboub, T., Luderer, G., Calvin, K., Demaere, G., Krey, V., Riahi, K., Rosler, H., Schaeffer, M., & Van Vuuren, D. P. (2013). What does the 2 C target imply for a global climate agreement in 2020? The LIMITS study on Durban platform scenarios. *Climate Change Economics, 4*(04), 1340008.

Leng, G. (2018). Keeping global warming within 1.5 °C reduces future risk of yield loss in the United States: A probabilistic modeling approach. *Science of the Total Environment, 644*, 52–59.

Lewis, J. I. (2009). Climate change and security: Examining China's challenges in a warming world. *International Affairs, 85*(6), 1195–1213.

Lewis, J. I. (2012). *Green innovation in China: China's wind power industry and the global transition to a low-carbon economy*. Columbia University Press.

Liu, L. C., Cheng, L., Zhao, L. T., Cao, Y., & Wang, C. (2020). Investigating the significant variation of coal consumption in China in 2002–2017. *Energy, 207*, 118307.

Lund, H., & Mathiesen, B. V. (2009). Energy system analysis of 100% renewable energy systems—The case of Denmark in years 2030 and 2050. *Energy, 34*(5), 524–531.

Magnan, A. K., Colombier, M., Billé, R., Joos, F., et al. (2016). Implications of the Paris agreement for the ocean. *Nature Climate Change, 6*(8), 732–735.

Mastini, R., Kallis, G., & Hickel, J. (2021). A Green New deal without growth? *Ecological Economics, 179*, 106832.

Maya-Drysdale, D., Jensen, L. K., & Mathiesen, B. V. (2020). Energy vision strategies for the EU green new deal: A case study of European Cities. *Energies, 13*(9), 2194.

Meckling, J., & Allan, B. B. (2020). The evolution of ideas in global climate policy. *Nature Climate Change, 10*(5), 434–438.

Meinshausen, M., Meinshausen, N., Hare, W., Raper, S. C., Frieler, K., Knutti, R., Frame, D. J., & Allen, M. R. (2009). Greenhouse-gas emission targets for limiting global warming to 2 C. *Nature, 458*(7242), 1158–1162.

Messner, D., Schellnhuber, J., Rahmstorf, S., & Klingenfeld, J. D. (2013). The Budget Approach—A Framework for a Global Transformation towards a Low Carbon Economy. In *Climate Change and Environmental Hazards Related to Shipping: An International Legal Framework* (pp. 9–33). Brill Nijhoff.

Michaelowa, A., Allen, M., & Sha, F. (2018). Policy instruments for limiting global temperature rise to 1.5 °C—Can humanity rise to the challenge? *Climate Policy, 18*(3), 275–286. https://doi.org/10.1080/14693062.2018.1426977

Millot, A., Krook-Riekkola, A., & Maïzi, N. (2020). Guiding the future energy transition to net-zero emissions: Lessons from exploring the differences between France and Sweden. *Energy Policy, 139*, 111358.

Minniti, M. (2013). *The dynamics of entrepreneurship: Evidence from global entrepreneurship monitor data*. OUP Oxford.

Mulugetta, Y., & Urban, F. (2010). Deliberating on low carbon development. *Energy Policy, 38*(12), 7546–7549.

New Climate Institute. (2020). *Navigating the nuances of net-zero targets*. Available from: https://newclimate.org/wp-content/uploads/2020/10/NewClimate_NetZeroReport_October2020.pdf

New, M., Liverman, D., Schroder, H., & Anderson, K. (2011). *Four degrees and beyond: The potential for a global temperature increase of four degrees and its implications*.

Nyambuu, U., & Semmler, W. (2020). Climate change and the transition to a low carbon economy–Carbon targets and the carbon budget. *Economic Modelling, 84*, 367–376.

Ou, Y. (2017). Amsterdam: Sustainable way to a green city. *QianXian, 4*, 74–79.

Pachauri, R. K., & Reisinger, A. (2007). IPCC fourth assessment report. *IPCC, Geneva, 2007*. Available from: https://archive.ipcc.ch/pdf/presentations/valencia-2007-11/pachauri-17-november-2007.pdf, presented in Valencia, Spain, 12–17 November 2007.

Park, C. E., Jeong, S. J., Joshi, M., Osborn, T. J., Ho, C. H., Piao, S., Chen, D., Liu, J., Yang, H., Park, H., Kim, B. M., & Feng, S. (2018). Keeping global warming within 1.5 C constrains emergence of aridification. *Nature Climate Change, 8*(1), 70–74.

Pettifor, A. (2020). *The case for the green new deal*. Verso.

Raftery, A. E., Zimmer, A., Frierson, D. M., Startz, R., & Liu, P. (2017). Less than 2 C warming by 2100 unlikely. *Nature Climate Change, 7*(9), 637–641.

Rajamani, L. (2010). The making and unmaking of the Copenhagen Accord. *International and Comparative Law Quarterly, 59*, 824.

Randalls, S. (2010). History of the 2 °C climate target. *Wiley Interdisciplinary Reviews: Climate Change, 1*(4), 598–605.

Raven, R., Sengers, F., Spaeth, P., Xie, L., Cheshmehzangi, A., & de Jong, M. (2019). Urban experimentation and institutional arrangements. *European Planning Studies, 27*(2), 258–281.

Rehfeldt, M., Fleiter, T., Herbst, A., & Eidelloth, S. (2020). Fuel switching as an option for medium-term emission reduction-A model-based analysis of reactions to price signals and regulatory action in German industry. *Energy Policy, 147*, 111889.

Roberts, C., Geels, F. W., Lockwood, M., Newell, P., Schmitz, H., Turnheim, B., & Jordan, A. (2018). The politics of accelerating low-carbon transitions: Towards a new research agenda. *Energy Research & Social Science, 44*, 304–311.

Rogelj, J., Luderer, G., Pietzcker, R. C., Kriegler, E., Schaeffer, M., Krey, V., & Riahi, K. (2015). Energy system transformations for limiting end-of-century warming to below 1.5 °C. *Nature Climate Change, 5*(6), 519–527.

Rogelj, J., Popp, A., Calvin, K. V., Luderer, G., Emmerling, J., Gernaat, D., Fujimori, S., Strefler, J., Hasegawa, T., Marangoni, G., & Krey, V. (2018). Scenarios towards limiting global mean temperature increase below 1.5 °C. *Nature Climate Change, 8*(4), 325–332.

Rodrigues, M., & Franco, M. (2018). Importance of living labs in urban Entrepreneurship: A Portuguese case study. *Journal of Cleaner Production, 180*, 780–789.

Sanderson, B. M., O'Neill, B. C., & Tebaldi, C. (2016). What would it take to achieve the Paris temperature targets? *Geophysical Research Letters, 43*(13), 7133–7142.

Schellnhuber, H. J., Rahmstorf, S., & Winkelmann, R. (2016). Why the right climate target was agreed in Paris. *Nature Climate Change, 6*(7), 649–653.

Shen, L., & Sun, Y. (2016). Review on carbon emissions, energy consumption and low-carbon economy in China from a perspective of global climate change. *Journal of Geographical Sciences, 26*(7), 855–870.

Skokic, V., & Morrison, A. (2011). Conceptions of tourism lifestyle entrepreneurship: Transition economy context. *Tourism Planning & Development, 8*(2), 157–169.

Solomon, S., Qin, D., Manning, M., Chen, Z., Marquis, M., Averyt, K., Tignor, M., & Miller, H. (2007). IPCC fourth assessment report (AR4). *Climate change, 374*. Available from: https://www.leg.mn.gov/docs/2015/other/150681/PFEISref_2/Solomon%20et%20al.%202007.pdf

Song, L., & Woo, W. T. (2008). *China's dilemma: Economic growth, the environment and climate change*. ANU Press.

Steininger, K. W., Meyer, L. H., Schleicher, S., Riahi, K., Williges, K., & Maczek, F. (2020). *Effort sharing among EU member states: Green deal emission reduction targets for 2030* (Short Version). Available from records of the EU action webpage at https://ec.europa.eu/clima/policies/effort_en#tab-0-0

Tanaka, K., & O'Neill, B. C. (2018). The Paris Agreement zero-emissions goal is not always consistent with the 1.5 °C and 2 °C temperature targets. *Nature Climate Change, 8*(4), 319–324.

Tang, X., Jin, Y., McLellan, B. C., Wang, J., & Li, S. (2018). China's coal consumption declining—Impermanent or permanent? *Resources, Conservation and Recycling, 129*, 307–313.

Tanneberger, F., Abel, S., Couwenberg, J., Dahms, T., Gaudig, G., Günther, A., Kreyling, J., Peters, J., Pongratz, J., & Joosten, H. (2021). Towards net zero CO_2 in 2050: An emission reduction pathway for organic soils in Germany. *Mires and Peat, 27*.

Taylor, M. A., Clarke, L. A., Centella, A., Bezanilla, A., Stephenson, T. S., Jones, J. J., Campbell, J. D., Vichot, A., & Charlery, J. (2018). Future Caribbean climates in a world of rising temperatures: The 1.5 vs 2.0 dilemma. *Journal of Climate, 31*(7), 2907–2926.

The United Nations Framework convention on Climate Change (UNFCCC). (2009). *Report of the Conference of the Parties on Its Fifteenth Session*, held in Copenhagen from 7–19 December 2009, Document Number: FCCC/CP/2009/11/Add.1, under the heading 'Decisions adopted by the Conference of the Parties', Available from: https://unfccc.int/resource/docs/2009/cop15/eng/11a01.pdf

The United Nations Framework convention on Climate Change (UNFCCC). (2010). *Cancun agreements.* Available from: https://unfccc.int/process/conferences/pastconferences/cancun-climate-change-conference-november-2010/statements-and-resources/Agreements

The United Nations Framework Convention on Climate Change (UNFCCC). (2019). *Climate action submit 2019*, Special event, held at UN headquarters in New York, the US, 23 Sep 2019, Available from: https://unfccc.int/event/climate-action-summit-2019

Tsiropoulos, I., Nijs, W., Tarvydas, D., & Ruiz, P. (2020). Towards net-zero emissions in the EU energy system by 2050. *Insights from scenarios in line with the 2030 and 2050 ambitions of the European Green Deal.*

United Nations. (2015). Paris agreement. In *Report of the Conference of the Parties to the United Nations Framework Convention on Climate Change (21st Session, 2015: Paris). Retrieved December* (Vol. 4, p. 2017). Cooperation Agreement is available from: https://www.unaoc.org/images/mou_iom.pdf

Urban, F. (2014). *Low carbon transitions for developing countries.* Routledge.

Vautard, R., Gobiet, A., Sobolowski, S., Kjellström, E., et al. (2014). The European climate under a 2 °C global warming. *Environmental Research Letters, 9*(3), 034006.

Wang, N., & Chang, Y. C. (2014). The development of policy instruments in supporting low-carbon governance in China. *Renewable and Sustainable Energy Reviews, 35*, 126–135.

Wang, X., Liu, C., Chen, S., Chen, L., Li, K., & Liu, N. (2020). Impact of coal sector's de-capacity policy on coal price. *Applied Energy, 265*, 114802.

Xie, L., Cheshmehzangi, A., Tan-Mullins, M., Flynn, A., & Heath, T. (2020). Urban entrepreneurialism and sustainable development: A comparative analysis of Chinese eco-developments. *Journal of Urban Technology, 27*(1), 3–26.

XinHuaNet. (2020). *China's commitment to be carbon neutral by 2060 is a key step in boosting global climate action, said Danish Minister for climate, energy and efficiency.* Retrieved February 19, 2021 from http://www.xinhuanet.com/2020-10/13/c_1126602137.htm

Yan, X., Ge, J., Lei, Y., & Duo, H. (2019). China's low-carbon economic transition: Provincial analysis from 2002 to 2012. *Science of the Total Environment, 650*, 1050–1061.

Zaklan, A., Wachsmuth, J., & Duscha, V. (2021). The EU ETS to 2030 and beyond: Adjusting the cap in light of the 1.5 °C target and current energy policies. *Climate Policy*, 1–14.

Zhang, L., Huang, Y. X., Li, Y. M., & Chen, X. L. (2010). An investigation on spatial changing pattern of CO_2 emissions in China. *Resources Science, 02*, 211–217.

Chapter 5
Decarbonised Race and New Destination in China

5.1 An Overview of Carbon Footprint in China by Available Data

The available data indicates China is by far the world's largest emitter of carbon dioxide. This is mainly because of China's large population and its rapid development and economic growth in recent decades. As China industrialised, its carbon dioxide emissions rose sharply, estimated from 3.1 billion tonnes in 2000 to 8.8 billion tonnes in 2012. Since 2012, the growth of China's CO_2 emissions has slowed due to the overall economic shift away from heavy and energy-intensive industries, slower economic growth, the switch from coal to gas, and the promotion of renewable energy (PRI, 2021). It is also important to note that China's per capita carbon dioxide emissions are lower than those of many industrialised countries. In 2018, China's per capita fossil fuel emissions were 7 tonnes, lower than the US's 16 tonnes and Japan's 9 tonnes. From 2005 to 2019, China's carbon dioxide emissions per unit of GDP were reduced by 48.1%. According to available data, carbon dioxide accounts for about 80% of China's national greenhouse gas emissions (UNFCCC, 2018). The International Energy Agency (IEA) estimates that China's energy-related carbon dioxide emissions under current policies will be more than three times of those under the Paris Agreement emission reduction path (IEA, 2019). According to the ICCSD (2020), China's emissions would need to peak by 2030 and then rapidly decline by 8–10% until 2050 to achieve emissions reductions in line with the Paris Agreement. Under such conditions, carbon dioxide emissions will be net-zero by 2050, and overall greenhouse gas emissions will be net-zero by 2060 (ICCSD, 2020).

The author compares the carbon emissions of different regions in China and obtains the results in Fig. 5.1. It can be seen that the eastern region of China has a relatively high level of low carbon economic development due to its apparent advantages in various aspects. But the energy demand is growing, especially in first-tier cities such as Beijing and Shanghai (Zhong, 2016). The central region, including several provinces in the north, has high carbon emissions and energy consumption. Although slightly better than the western region, the figures are still behind the eastern

A. Cheshmehzangi and H. Chen, *China's Sustainability Transitions*,
https://doi.org/10.1007/978-981-16-2621-0_5

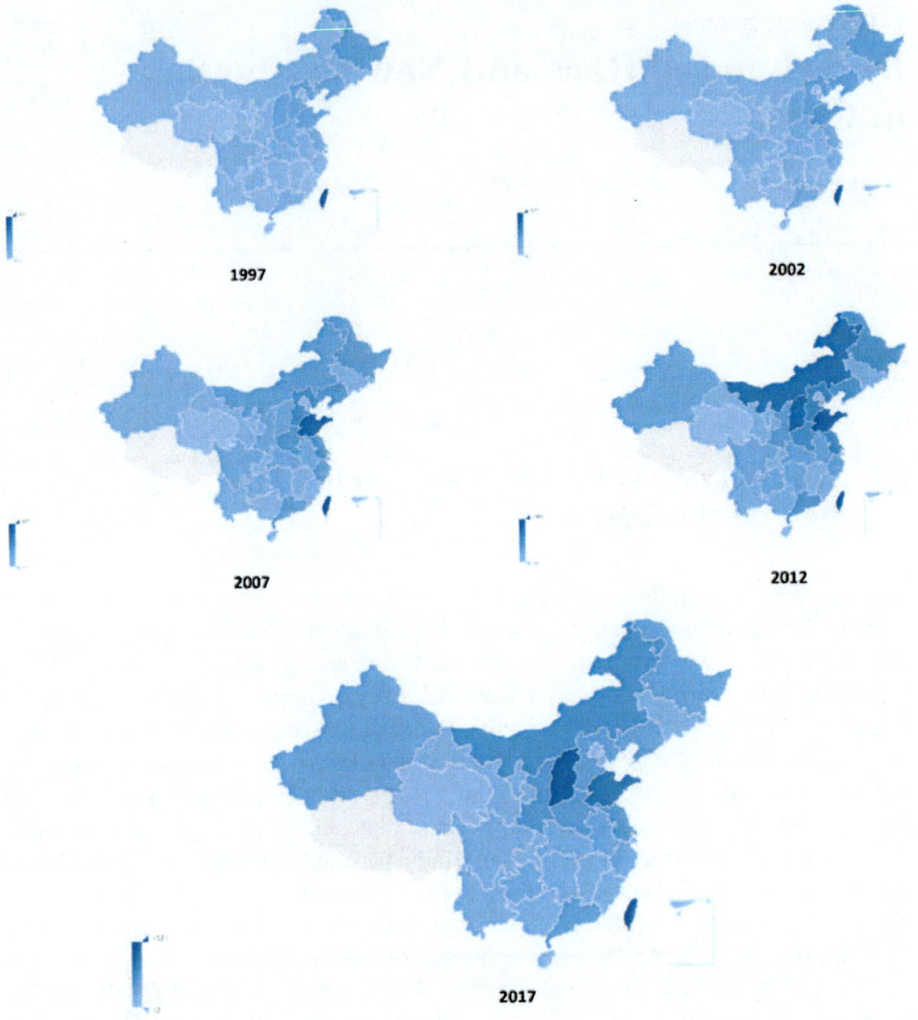

Fig. 5.1 Distribution of CO_2 emission (mt) in China (Drawn by the authors, data resource from: www.ceads.net)

region. Economic development in west China is relatively backward, and it relies more on the input of energy resources, but the energy utilisation efficiency is low, so the carbon emission in some areas is high (Zhang et al., 2010).

China's current economic growth is significantly dependent on energy consumption and carbon emissions. From the primary industry perspective, the mechanism of low carbon agricultural products has not yet been formed (As shown in Fig. 5.2).

Given the secondary industry, China's industrial sector is characterised by significantly high carbon (Su et al., 2012). In transportation, energy conservation and emission reduction of scientific and technological innovation and promotion efforts are insufficient, lack of investment in research and development, new energy vehicles

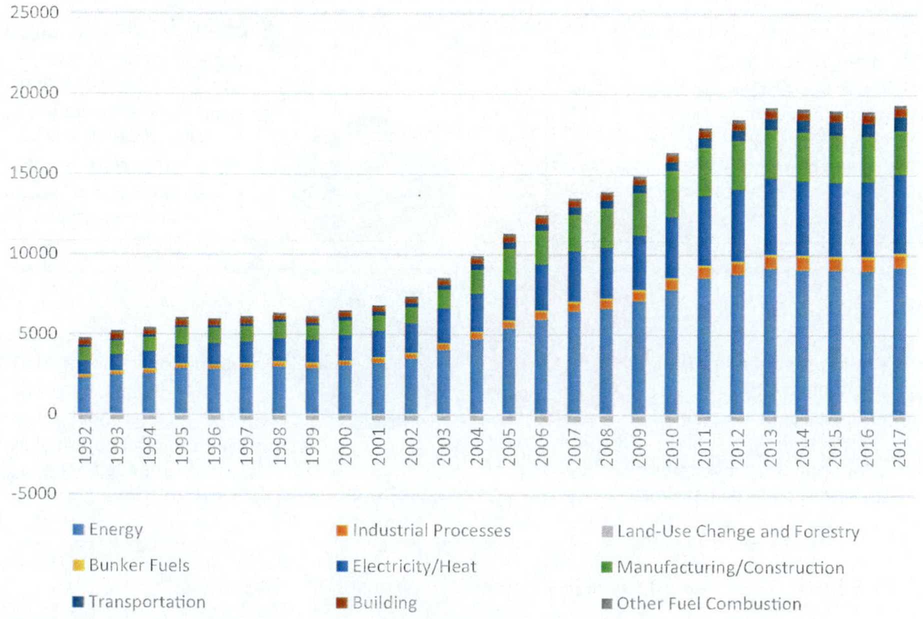

Fig. 5.2 Distribution of CO_2 emission (mt) in different industries of China (Drawn by the authors, data resource from: https://www.climatewatchdata.org/)

are still in the stage of development. China's energy structure continues to improve, and the overall energy structure is improving. However, the proportion of coal is still very high (Fang et al., 2016). The situation of low carbon consumption is still not optimistic. Consumption habits bring high carbon consumption, and low carbon products are sold at high prices and difficult to be popularised. Based on the above conclusions, the author summarises the current situation, advantages, and disadvantages of China's low carbon development and its future prospects, as shown in Fig. 5.3 (Kang et al., 2018).

5.2 China's Low Carbon Target: The Commitment and Momentous Goal

Over the past four decades, China has made remarkable economic development achievements (Brandt & Rawski, 2008; Cheshmehzangi, 2020; Pak & Park, 1997; Sinton et al., 1998; Yue, 2012; Zhou et al., 2017). Its GDP has proliferated, urbanisation has accelerated, and the number of poor people has been significantly reduced. However, the traditional extensive economic growth model has also led to the doubling of resource and energy consumption, pollutants, and carbon emissions, which have brought substantial environmental and climate risks to society. China's total energy consumption in 2019 was 4.86 billion tons of standard coal, and China's

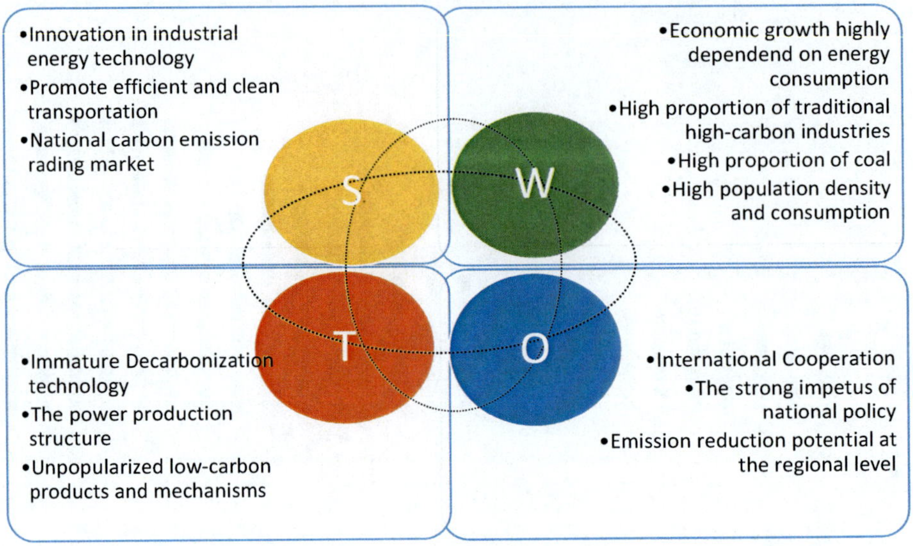

Fig. 5.3 SWOT analysis of China low carbon development (Drawn by the authors)

total energy-related carbon dioxide emissions in 2019 reached 9.8 billion tons, ranking the first in the world (NBS, 2020). In the future, China's economy is in a transition period from rapid development to high-quality development. This initiative is highlighted in China's high-quality urbanisation (Cheshmehzangi, 2016), to respond to some of the ongoing challenges (Guan et al., 2018), consider regional habitat quality (Bai et al. 2019), specific environmental quality enhancement (Ren et al., 2014), and reverse some of the losses (Qiu et al., 2020). More importantly, China has committed to act more responsibly to ensure low carbon targets are implemented and met in a transitional process. For instance, China's climate policy towards reaching a carbon peak in 2030 (Liu et al., 2015) indicates opportunities for restructuring and revising some existing institutions. Therefore, there is room for green innovation (Lewis, 2012), as well as for low carbon pathways that suggest high-level improvements in air quality (Li et al., 2019), the restarting of the national electricity system (Kahrl et al., 2011), and achieving environmental co-benefits (Yang et al., 2021). By comparing China and the global models (Chen et al., 2016), we see differences and similarities; but more importantly, context-specific challenges (Cheshmehzangi and Dawodu, 2019a), which are the backbone of industrial transformation (Liu et al., 2016), changing lifestyles (Hubacek et al., 2012), and a move towards low carbon economy (Zhang, 2010). These are challenging not only for China but also for any country in such a position.

With 29% of the world's carbon dioxide emissions in 2019, China faces the task of reducing its emissions by 2060 when it aims to be "carbon neutral". Therefore, it is significant to speed up building a clean, low carbon, safe and efficient energy system. On September 22, 2020, President Xi Jinping pledged to the world that China's carbon dioxide emissions would peak by 2030 and that China would strive to become carbon neutral by 2060. Thus, the next four decades represent a long process of achieving the ultimate goal of carbon neutrality, which will be adopted and replicated by countries looking up to China as a successful model. At the Climate Ambition Summit, President Xi stressed that China had made important contributions to the conclusion of the Paris Agreement on tackling climate change and is an active practitioner of the implementation (Calliari et al., 2020). This was followed by earlier attempts to highlight the global climate action plans (Arroyo, 2018; Streck, 2020) and consider a shift towards achieving the sustainable development goals (SDGs) (Malpass, 2020). President Xi referred to some further commitments for 2030 at the 2020 Climate Ambition Summit (Global Times, 2020):

> *China will lower its carbon dioxide emissions per unit of GDP by over 65 percent from the 2005 level, increase the share of non-fossil fuels in primary energy consumption to around 25 percent, increase the forest stock volume by 6 billion cubic meters from the 2005 level, and bring its total installed capacity of wind and solar power to over 1.2 billion kilowatts.*

Figure 5.4 summarises some of the key target nodes of the recent past and the coming decades, which are set by the Chinese government. While we see gradual progress, we also see critical targets throughout the whole plan. To make them realistically happen, a step-by-step or systematic progression is needed. The plan, therefore, defines China's low carbon transition, which is set to be a long and challenging journey.

5.3 A Brief Overview of the Sustainable Development Goals (SDGs)

China's sustainable development strategies are identified through earlier phases or plans for sustainable energy use (Li & Oberheitmann, 2009), energy situation strategy (Zhang et al., 2011), environmental protection strategies (Zhang & Wen, 2008), China's energy demand (Ahmed & Ozturk, 2018), etc. These are all before the birth of the Sustainable Development Goals (SDGs). In September 2015, the UN Summit on Sustainable Development adopted the 'Transforming our World: 2030 Agenda for Sustainable Development' agreed by the 193 member states of the United Nations (General Assembly, 2015; United Nations, 2015). The Agenda is another guiding document on the global development process, which contains 17 SDGs, 169 targets, and 232 indicators (Cheshmehzangi and Dawodu, 2019b; Paoli & Addeo, 2019). It cuts across the economic, social, and environmental dimensions and provides a new roadmap and indicator for global development. The correlation between the SDGs and low carbon pathways is increasingly seen in various ways of green investing (Ibragimov et al., 2019), assessing energy projects (Castor et al., 2020), climate change plans (Gomez-Echeverri, 2018), and global warming reductions (Von Stechow et al., 2016). By realising the transformative potentials of the SDGs, as highlighted by Elder and King (2018), we see a broad opportunity, which also includes specificities of low carbon transitions.

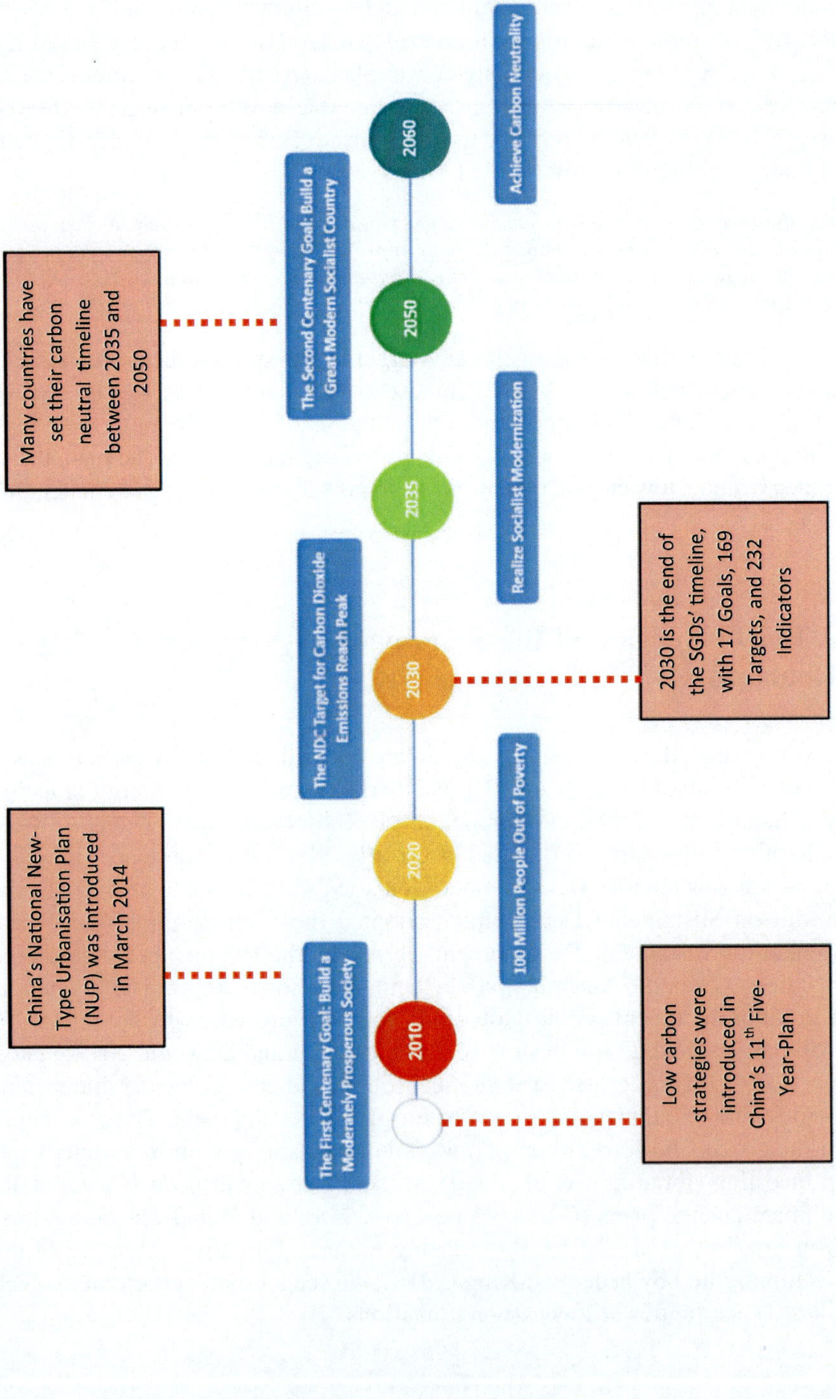

Fig. 5.4 Key target nodes set by Chinese government (Drawn by author)

As the largest developing country in the world, China has always given top priority to development and ecological and environmental protection. During the recently-completed 13th Five-Year Plan period, China has made sustainable development and ecological progress its basic state policy. In order to implement the SDGs, the Chinese government has taken a series of actions and made positive progress in top-level design, strategic alignment, institutional safeguards, international exchanges, and South-South cooperation (Ali et al., 2018; Bergamaschi et al., 2017; Besharati, 2019; Chen et al., 2019; Corrêa, 2017; Li et al., 2018; Liu et al., 2019; Weigel and Demissie, 2021; Xie et al., 2021; Xue et al., 2018; Zhu et al. 2019). China has established a domestic coordination mechanism for implementing the 2030 Agenda, led by the Ministry of Foreign Affairs, with 43 government departments cooperating with each other (Cheshmehzangi et al., 2021; Li & Zhu, 2019; Zhu, 2017). In March 2017, the China International Development Knowledge Centre (CIKD) was officially approved to be established, providing a platform for China and other countries to study and exchange development theories and practices related to the 2030 Agenda (CIKD, n.d.). One of China's key aspects of the recent development has been to actively implement the 2030 Agenda for Sustainable Development. It has been included in Chapter 53 of the Outline of the 13th Five-year Plan for National Economic and Social Development of the People's Republic of China to actively assume international responsibilities and obligations. There are traces of this implementation plan in various parts of the globe, through project partnerships, and specifically through the Belt and Road Initiative (BRI) (Dong et al., 2018; Li & Zhu, 2019; Teo et al., 2019).

In recent years, the Chinese government has taken a series of actions regarding top-level design, strategic alignment, and institutional guarantee (Cheshmehzangi et al., 2021; Li & Lin, 2017; Zhang et al., 2015). The SDGs have been incorporated into the 13th Five-Year Plan and the national medium and long-term overall development plan. This integration is multifaceted and is reflected in the prevailing social, economic and environmental planning. The 2030 Agenda covers economic development, social progress, and environmental protection. In the economic field, for example, the Chinese government has formulated the "Outline of the National Strategy for Innovation-Driven Development" and the "National Plan for Sustainable Agricultural Development (2015–2030)". The former was proposed for implementation in the 18th CPC National Congress (Central Committee of the Communist Party of China (CPC) and the PRC State Council, 2016), and the latter is a long-term plan by the PRC State Council (2015a). In the social field, the "Decision on Winning the Battle against Poverty" by PRC State Council (2015b) and the "Outline of the Plan for a Healthy China" by PRC State Council (2016a) were issued. In the environmental field, China has formulated the "National Biodiversity Conservation Strategy and Action Plan (2011–2030)" by MEE (Ministry of Ecology and Environment of PRC, 2010) and the "National Climate Change Plan (2014–2020)" by the PRC State Council (2014a). "No Poverty" is the first sustainable development goal, and the country's 13th Five-Year Plan has set a goal of eradicating poverty by 2020, launching "The decision to Win the Battle Against Poverty" by PRC State Council

(2016b). China has introduced and implemented a series of relevant policies that are in line with the SDGs (see Table 5.1).

In this regard, we see China's commitment to implementing and achieving SDGs is tangible. For instance, there are examples of institutional arrangement (and rearrangement), which are aimed to strengthen China's position in achieving the SDGs (Mao et al., 2019). Besides, other major initiatives and mechanisms play a significant part in China's plan for achieving its targets based on the SDGs. These are

Table 5.1 China's major policies corresponding to the sustainable development goals during the 13th five-year plan period (Collected from Internet by author)

Sustainable development goals	Policies issued by China or participates in
SDGs1: No poverty	The decision to win the battle against poverty
SDGs2: Zero hunger	National plan for sustainable agricultural development (2015–2030)
SDGs3: Good health and wellbeing	Outline of the "healthy China" program
SDGs4: Quality education	Outline of the national Mid- to long-term plan for education reform and development (2010–2020)
SDGs5: Gender equality	Program for the development of Chinese women
SDGs6: Clean water and sanitation	Action plan for water pollution control
SDGs7: Affordable and clean energy	Strategic action plan for energy development (2014–2020)
SDGs8: Decent work and economic growth)	The Ten-year framework for the programme on sustainable consumption and production patterns
SDGs9: Industry, innovation and infrastructure	Outline of the national strategy for innovation-driven development
SDGs10: Reduced inequalities	Promote the development planning of ethnic areas and ethnic groups with small population
SDGs11: Sustainable cities and communities	Urban and rural community service system construction plan (2016, 2020)
SDGs12: Sustainable consumption and production	Opinions on speeding up the establishment of green production and consumption regulations and policy system
SDGs13: Climate action	Work plan for controlling greenhouse gas emissions during the 13th five-year plan period
SDGs14: Life under water	Provisions on the administration of fishing permits
SDGs15: Life on land	The Outline of the 13th Five-Year Plan for ecological protection
SDGs16: Institutions, good governance	Opinions of The State Council on strengthening and improving community governance in urban and rural areas
SDGs17: Partnerships for the goals	Addis Ababa Action Agenda

summarised as policy instruments (Xie et al., 2021), development of co-targets for the climate policy, particularly in the regional and sectoral aspects (Zhang et al., 2020), and possibilities for trade-offs between the SDGs (Zhang et al., 2019). In all cases, there are strong examples of the role and functionality of policies utilised for adjusting or optimising the status and pathways that China has initiated in its recent five-year-plans (FYPs). To sum, we can say the battle against climate change and its impacts is just started.

5.4 Combatting Climate Change Through Low Carbon Targets

China has taken emission reduction measures to integrate the low carbon concept into all aspects and processes of social and economic. At present, China has integrated global climate change into its overall strategy of economic and social development. In terms of the international response to and governance of climate change, China was one of the first countries to sign the United Nations Framework Convention on Climate Change (1992) and Kyoto Protocol (1997) (Zhuang, 2007) and was among the first developing countries to formulate and implement a national program to address global climate change (XinHuaNet, 2015). In June 2007, the Chinese government released the "National Plan on Climate Change" (State Council, 2007), which fully elaborated the major policy measures to address global climate change by 2010. It is not only China's first comprehensive policy document on global climate change but also the first national program of a developing country in this field. In October 2008, the Chinese government issued a white paper entitled "China's policies and actions on climate change" (State Council, 2008)., which thoroughly explains China's major policies and actions to mitigate and adapt to global climate change. In 2009, the Chinese government took an active part in the United Nations Climate Change Conference held in Copenhagen and promised to reduce carbon dioxide emissions per unit of GDP by 40–45% by 2020 compared with 2005 (Liu et al., 2010).

In November 2013, China released "Adapt to Climate Change Strategy" (NDRC, 2013), the first strategic plan specifically aimed at adapting to global climate change. In December 2015, the Chinese government took part in the United Nations Climate Change Conference in Paris and promoted the Paris Agreement (Chao et al., 2016). According to the agreement, China's Intended Nationally Determined Contributions (INDCs) include peaking carbon dioxide emissions by 2030 and striving to achieve them as soon as possible, reducing carbon dioxide emissions per unit of GDP by 60–65% from 2005 levels (Huo et al., 2016), increasing the share of non-fossil energy to around 20%, and achieving a forest carbon sink of 4.5 billion cubic meters (Qu et al., 2016). Simultaneously, the Chinese government has promoted "South-South cooperation" in recent years and provided support to countries and regions with lower levels of development. From 2011 to 2014, China allocated a total of 270 million yuan to help developing countries improve their capacity to cope with global climate change. The first steps are already taken in the process of international cooperation, and some with the least developing countries or poorer nations in the region.

At the domestic level of response and governance, China attaches great importance to the formulation and implementation of a national strategy to address climate change. Besides, China has launched "National Plan on Climate Change" by PRC State Council (2007), "White paper: China's policies and actions on climate change" by PRC State Council (2008), "Adapt to Climate Change Strategy" (NDRC, 2013). Moreover, the Chinese government has promulgated a series of special or related planning and scheme of response to global climate change, such as the "12th Five-Year Control Scheme for Greenhouse Gas Emissions" by PRC State Council (2011). It also published "Action Plan for Energy Conservation and Emissions Reduction 2014–2015" by PRC State Council (2014b) and the "National Plan on Climate Change (2014, 2020)" (NDRC, 2014), etc. (see Table 5.2). As shown here, we can see a progressive attempt by China since the 11th FYP, with many new policy documents, reports, outlines, and low carbon agenda.

5.5 China's New Commitment and Its Significance: The Carbon Neutrality Goal by 2060

The recent commitment to the carbon neutrality goal is a pivotal moment for China. This is especially more important when the county (like many other countries) is currently facing national and international upheaval from the global COVID-19 pandemic. To achieve President Xi's 2060 carbon–neutral target, China must now propose a more resilient, low carbon, and sustainable medium to long-term economic reconstruction plan. In addition, it must update its Nationally Determined Contribution (NDC) to align with the 2060 target as well as its detailed low carbon development strategy. The 14th Five-Year Plan (FYP) has been released recently and is expected to impose stricter constraints on energy consumption and emissions. At the same time, it is more urgent to build clean, low carbon, safe, and efficient energy systems. These are ongoing strategies and plans that are even more important in this current decade and before the 2030 targets. The path afterward and until 2060 is very much dependant on what will happen during the timeline of the 14th and 15th FYPs. When China's carbon peak is reached in 2030, the SDGs timeline will be over, too. In reality, the process will require careful development and promotion of infrastructures, capacities, capabilities, technologies, policies, benchmarks, and institutional arrangements.

China's energy and resource endowments and current national conditions have determined that it is impossible to develop a single energy source. Instead, China must complement in multiple ways and combine coal, oil, natural gas, hydropower, wind power, solar energy, biomass energy, and nuclear energy. It is estimated that before reaching "carbon peak" in the next 10 years, China will still be dominated by fossil

Table 5.2 China's climate change policies and main contents (Collected by the authors)

Policy document	Year	Main content
Implementation of low carbon Pilot projects in Provinces and Cities	2010	Pilot projects will be carried out in Guangdong, Liaoning, Hubei, Shaanxi and Yunnan provinces, and in Tianjin, Chongqing, Shenzhen, Xiamen, Hangzhou, Nanchang, Guiyang and Baoding
Action plan for air pollution prevention and control	2013	Further control total coal consumption, accelerate the goals and requirements of replacing clean energy with clean energy, and substantially increase our efforts to control fossil fuel consumption and develop clean energy
Interim measures for the administration of carbon emission trading	2014	Clear the national carbon market construction ideas. Guidelines on GHG emission accounting and reporting for 24 key industries were issued, and a direct reporting system for GHG emission data of enterprises was established
National climate change plan (2014–2020)	2014	Set out the main targets and priorities for addressing climate change by 2020
Strengthening action on climate change—China's Intended Nationally Determined Contribution(INDCs)	2015	Clear target CO_2 emissions to peak around 2030
Work plan for controlling greenhouse gas emissions during the 13th Five-Year Plan Period	2016	Promote green and low carbon development and ensure that we meet the low carbon development targets set out in the 13th Five-Year Plan
Industrial green development plan (2016–2020)	2016	Promote the establishment of a green manufacturing system and accelerate the transformation of industry to low carbon growth
National carbon emission trading market construction scheme (power generation industry)	2017	The national carbon market construction task will be deployed, and the national carbon market will be established in stages and step by step with the power generation industry as the breakthrough point
National key energy conservation and low carbon technology promotion directory	2017	Covers non-fossil energy, fuel and raw material substitution, industrial processes and other non-carbon dioxide emission reduction, carbon capture, utilisation and storage, carbon sink and other fields, and a total of 27 low carbon technologies that are mainly promoted by the state

energy. Therefore, it is necessary to realise clean utilisation of fossil energy, promote a high proportion of wind power and photovoltaic application, and make joint efforts to achieve the goal of "carbon neutral". At the national level, if China implements a profound path of emission reduction under intensified actions, it will not only significantly reduce greenhouse gas emissions and contribute to the achievement of the Paris Agreement temperature control targets but also avoid natural disasters caused by climate change, such as mitigating sea level rise and water shortage. In doing so, China could potentially generate direct climate benefits. At the same time, controlling carbon emissions will improve air quality, improve the living environment and promote human health. Undoubtedly, this will be beneficial to the health of the economy, society, and the environment. Compared with the existing policy path, choosing the path of profound emission reduction will gradually reduce its emission reduction cost (including equipment investment, operation, and maintenance cost, etc.) in the later period. This could ultimately help to achieve positive economic benefits during the current 14th Five-Year Plan period. Therefore, this path will bring sustainable benefits/services to China's economic and social development in the next 30 years. In doing so, it could effectively promote the win–win situation of climate and economic development.

For example, as the only national renewable energy demonstration zone approved by The State Council, Zhangjiakou has taken the lead in becoming a "negative carbon" city. Zhangjiakou is rich in renewable energy and the installed wind power in the national prefecture-level city ahead of the ten million kW installed mark (Zhangjiakou News, 2020). The local government is actively promoting hydrogen production from wind power and other renewable energy sources, which is thriving in Zhangjiakou. Key factors such as 'relying on the abundant renewable energy sources such as wind power in Zhangjiakou', 'producing green hydrogen', and 'producing methanol with green hydrogen and carbon dioxide' can expand the application industry chain of hydrogen energy. In doing so, they can reduce carbon emissions, realise multiple benefits such as carbon utilisation, and help achieve 'carbon peak' and 'carbon neutral' as soon as possible (XinHuaNet, 2020). In the following chapters, we will delve into more examples, some that China could learn from and some that we value as experimentation cases for the forthcoming steps. As we close this chapter, we highlight the importance of 'transition', which becomes more realistic and feasible in achieving the sustainability goals. This is regardless of the SDGs or any other sort of targets that China (or any other country) may have an eye on. Thus, we must demonstrate what progress has happened so far, what lessons we can gather, and what pathways are ahead. In doing so, not only we can explore China's microhistory of low carbon development but also its transition towards the carbon neutrality goal of 2060.

References

Ahmed, K., & Ozturk, I. (2018). What new technology means for the energy demand in China? A sustainable development perspective. *Environmental Science and Pollution Research, 25*(29), 29766–29771.

Ali, S., Hussain, T., Zhang, G., Nurunnabi, M., & Li, B. (2018). The implementation of sustainable development goals in "BRICS" countries. *Sustainability, 10*(7), 2513.

Arroyo, V. (2018). The global climate action summit: increasing ambition during turbulent times. *Climate Policy, 18*(9), 1087–1093.

Bai, L., Xiu, C., Feng, X., & Liu, D. (2019). Influence of urbanization on regional habitat quality: A case study of Changchun City. *Habitat International, 93*, 102042.

Bergamaschi, I., Moore, P., & Tickner, A. B. (Eds.). (2017). *South-South cooperation beyond the myths: Rising donors, new aid practices?* Springer.

Besharati, N. A. (2019). Measuring effectiveness of South-South cooperation. *Occasional Paper*, (52).

Brandt, L., & Rawski, T. G. (Eds.). (2008). *China's great economic transformation.* Cambridge University Press.

Calliari, E., Mysiak, J., & Vanhala, L. (2020). A digital climate summit to maintain Paris Agreement ambition. *Nature Climate Change, 10*(6), 480–480.

Castor, J., Bacha, K., & Nerini, F. F. (2020). SDGs in action: A novel framework for assessing energy projects against the sustainable development goals. *Energy Research & Social Science, 68*, 101556.

Central Committee of the Communist Party of China (CPC) and the PRC State Council. (2016). Outline of the National innovation-driven development strategy, issued by the CPC Central Committee and the State Council <中共中央国务院印发 (国家创新驱动发展战略纲要)>. Translated in 2019, by the Center for Security and Emerging Technology (CSET). Available from: https://cset.georgetown.edu/wp-content/uploads/t0076_innovation_driven_development_strategy_EN.pdf.

Chao, Q. C., Zhang, Y. X., Gao, X., & Wang, M. (2016). Paris agreement: A new starting point for global climate governance. *Advances in Climate Change Research,* (1).

Chen, W., Yin, X., & Zhang, H. (2016). Towards low carbon development in China: A comparison of national and global models. *Climatic Change, 136*(1), 95–108.

Chen, S., Guo, L., Wang, Z., Mao, W., Ge, Y., Ying, X., & Tang, S. (2019). Current situation and progress toward the 2030 health-related sustainable development goals in China: A systematic analysis. *PLoS Medicine, 16*(11), e1002975.

Cheshmehzangi, A. (2016). China's New-type Urbanisation Plan (NUP) and the foreseeing challenges for decarbonization of cities: A review. *Energy Procedia, 104*, 146–152.

Cheshmehzangi, A. (2020). Evaluating the Nexus between housing and energy sectors: the comparison of urban, Peri-urban and rural housing areas in Zhuhai, China. *Energy and Power Engineering, 12*(06), 314.

Cheshmehzangi, A., & Dawodu, A. (2019a). *Sustainable urban development in the age of climate change – People: Cure or curse* (pp. 133–156). Palgrave Macmillan.

Cheshmehzangi, A., & Dawodu, A. (2019b). The review of Sustainable Development Goals (SDGs): People, perspective and planning. In *Sustainable urban development in the age of climate change – People: Cure or curse* (pp. 133–156). Palgrave Macmillan.

Cheshmehzangi, A., Xie, L., & Tan-Mullins, M. (2018). The role of international actors in low-carbon transitions of Shenzhen's international low carbon city in China. *Cities, 74*, 64–74.

Cheshmehzangi, A., Dawodu, A., & Sharifi, A. (2021). *Sustainable urbanism in China.* Routledge.

Corrêa, M. L. (2017). Quantification of South-South cooperation and its implications to the foreign policy of developing countries. *South Centre Policy Brief, 41.*

Dong, L., Yang, X., & Li, H. (2018). The belt and road initiative and the 2030 agenda for sustainable development: Seeking linkages for global environmental governance. *Chinese Journal of Population Resources and Environment, 16*(3), 203–210.

Elder, M., & King, P. (2018). Realising the transformative potential of the SDGs.

Fang, C. L., Wang, S. J., & Wang, Y. (2016). Current situation and countermeasures of Chinese low-carbon eco-metro area development. *Geographical Research, 09*, 1601–1614.

General Assembly (2015). Sustainable development goals. SDGs transform our world, 2030. *Global Change*, Issue 82, 20–23. Available from: http://www.igbp.net/download/18.62dc35801456272b46d51/1399290813740/NL82-SDGs.pdf.

Global Times. (2020). President Xi Jinping delivers an important speech at the 2020 Climate Ambition Summit on Saturday. Available from: https://www.globaltimes.cn/content/1209756.shtml.

Gomez-Echeverri, L. (2018). Climate and development: Enhancing impact through stronger linkages in the implementation of the Paris Agreement and the Sustainable Development Goals (SDGs). *Philosophical Transactions of the Royal Society A: Mathematical, Physical and Engineering Sciences, 376*(2119), 20160444.

Guan, X., Wei, H., Lu, S., Dai, Q., & Su, H. (2018). Assessment on the urbanization strategy in China: Achievements, challenges and reflections. *Habitat International, 71*, 97–109.

Hubacek, K., Feng, K., & Chen, B. (2012). Changing lifestyles towards a low carbon economy: An IPAT analysis for China. *Energies, 5*(1), 22–31.

Huo, J., Weng, Y. Y., & Zhang, X. L. (2016). Analysis of China's low-carbon energy economy transition path in 2050. *Environmental Protection, 16*, 38–42.

ICCSD (Institute of Climate Change and Sustainable Development, Tsinghua University). (2020). China low-carbon development strategy and transformation pathways.

International Energy Agency (IEA). (2019). World Energy Outlook 2019. Retrieved 21 January 2021 from https://www.iea.org/events/world-energy-outlook-2019.

Kahrl, F., Williams, J., Jianhua, D., & Junfeng, H. (2011). Challenges to China's transition to a low carbon electricity system. *Energy Policy, 39*(7), 4032–4041.

Kang, Z. Y., Li, K., & Qu, J. (2018). The path of technological progress for China's low-carbon development: Evidence from three urban agglomerations. *Journal of Cleaner Production, 178*, 644–654.

Lewis, J. I. (2012). *Green innovation in China: China's wind power industry and the global transition to a low-carbon economy*. Columbia University Press.

Li, J., & Lin, X. (2017). Assessing credit guarantee companies in China: Applying a new framework. *China Economic Review, 44*, 98–111.

Li, Y., & Oberheitmann, A. (2009). Challenges of rapid economic growth in China: Reconciling sustainable energy use, environmental stewardship and social development. *Energy Policy, 37*(4), 1412–1422.

Li, Y., & Zhu, X. (2019). The 2030 agenda for sustainable development and China's belt and road initiative in Latin America and the Caribbean. *Sustainability, 11*(8), 2297.

Li, L., Xia, X. H., Chen, B., & Sun, L. (2018). Public participation in achieving sustainable development goals in China: Evidence from the practice of air pollution control. *Journal of Cleaner Production, 201*, 499–506.

Li, N., Chen, W., Rafaj, P., Kiesewetter, G., Schöpp, W., Wang, H., Zhang, H., Krey, V., & Riahi, K. (2019). Air quality improvement co-benefits of low-carbon pathways toward well below the 2 °C climate target in China. *Environmental Science & Technology, 53*(10), 5576–5584.

Liu, W. D., Lei, Z., Wang, L. M., Zhao, J. A., Ma, L., & Tang, Z. P. (2010). A preliminary study on the development framework of low-carbon economy in China. *Geographical Research, 29*(5), 778–788.

Liu, Z., Guan, D., Moore, S., Lee, H., Su, J., & Zhang, Q. (2015). Climate policy: Steps to China's carbon peak. *Nature News, 522*(7556), 279.

Liu, G., Hao, Y., Zhou, Y., Yang, Z., Zhang, Y., & Su, M. (2016). China's low-carbon industrial transformation assessment based on Logarithmic Mean Divisia Index model. *Resources, Conservation and Recycling, 108*, 156–170.

Liu, J. Y., Fujimori, S., Takahashi, K., Hasegawa, T., Wu, W., Takakura, J. Y., & Masui, T. (2019). Identifying trade-offs and co-benefits of climate policies in China to align policies with SDGs and achieve the 2 °C goal. *Environmental Research Letters, 14*(12), 124070.

Malpass, D. (2020). Remarks at the climate ambition summit 2020. Speeches of World Bank Presidents, Published by World Bank. https://doi.org/10.1596/34933

Mao, Z., Xue, X., Tian, H., & Michael, A. U. (2019). How will China realize SDG 14 by 2030?—A case study of an institutional approach to achieve proper control of coastal water pollution. *Journal of Environmental Management, 230*, 53–62.

MEE (Ministry of Ecology and Environment of the People's Republic of China). (2010). National Biodiversity Conservation Strategy and Action Plan (2011–2030) <中国生物多样性保护战略与行动计划>. Available from: http://www.mee.gov.cn/gkml/hbb/bwj/201009/t20100921_194841.htm.

NBS (National Bureau of Standards). (2020). Statistical bulletin of the People's Republic of China on National economic and social development 2019 <中华人民共和国2019年国民经济和社会发展统计公报>. Available from: http://www.stats.gov.cn/tjsj/zxfb/202002/t20200228_1728913.html.

NDRC (National Development and Reform Commission). (2013). Adapt to climate change strategy <国家适应气候变化战略>. Available from: https://www.ndrc.gov.cn/xxgk/zcfb/tz/201312/t20131209_963985.html.

Pak, C. D., & Park, J. D. (1997). *The special economic zones of China and their impact on its economic development*. Greenwood Publishing Group.

Paoli, A. D., & Addeo, F. (2019). Assessing SDGs: A methodology to measure sustainability. *Athens Journal of Social Sciences, 6*, 229–250.

Principles for Responsible Investment (PRI). (2021). *Delivering carbon neutrality in China*. Retrieved 21 January 2021 from https://www.unpri.org/pri-blogs/delivering-carbon-neutrality-in-china/7000.article.

Qiu, B., Li, H., Tang, Z., Chen, C., & Berry, J. (2020). How cropland losses shaped by unbalanced urbanization process? *Land Use Policy, 96*, 104715.

Qu, C., & Chen, T. (2016). Estimation of China's carbon emission intensity reduction potential in 2030. *China Population, Resources and Environment, 26*(007), 62–69.

Ren, L., Cui, E., & Sun, H. (2014). Temporal and spatial variations in the relationship between urbanization and water quality. *Environmental Science and Pollution Research, 21*(23), 13646–13655.

Sinton, J. E., Levine, M. D., & Qingyi, W. (1998). Energy efficiency in China: Accomplishments and challenges. *Energy Policy, 26*(11), 813–829.

State Council. (2007). National plan on climate change <中国应对气候变化国家方案>. Available from: http://www.gov.cn/gongbao/content/2007/content_678918.htm.

State Council. (2008). White paper: China's policies and actions on climate change <中国应对气候变化的政策与行动>. Available from: http://www.scio.gov.cn/zfbps/ndhf/2008/Document/307869/307869.htm.

State Council. (2011). 12th Five-Year control scheme for greenhouse gas emissions <"十二五"控制温室气体排放工作方案>. Available from: http://www.gov.cn/zhengce/content/2012-01/13/content_1294.htm.

State Council. (2014a). National climate change plan (2014–2020) <国家应对气候变化规划>. Available from: http://www.scio.gov.cn/xwfbh/xwbfbh/wqfbh/2015/20151119/xgzc33810/document/1455885/1455885.htm.

State Council. (2014b). Action plan for energy conservation and emissions reduction (2014–2015) <节能减排低碳发展行动方案>. Available from: http://www.gov.cn/zhengce/content/2014-05/26/content_8824.htm.

State Council. (2015a). National sustainable agriculture development plan (2015–2030) <全国农业可持续发展规划>. Available from: https://www.chinawaterrisk.org/research-reports/national-sustainable-agriculture-development-plan-2015-2030/.

State Council (2015b). Decision on winning the battle against poverty <关于打赢脱贫攻坚战的决定>. Available from: http://www.gov.cn/zhengce/2015-12/07/content_5020963.htm.

State Council (2016a). Outline of the plan for a healthy China. <"健康中国2030"规划纲要>. Available from: http://www.gov.cn/xinwen/2016-10/25/content_5124174.htm.

State Council. (2016b). The 13th five-year plan for economic and social development of the People's Republic of China <国民经济和社会发展第十三个五年规划>. Available from: http://www.gov.cn/xinwen/2016-03/17/content_5054992.htm.

Streck, C. (2020). The mirage of Madrid: elusive ambition on the horizon. *Climate Policy, 20*(2), 143–148.

Su, M. R., Chen, B., Chen, C., Yang, Z. F., Liang, C., & Wang, J. (2012). Reflection on upsurge of low-carbon cities in China: Status quo, problems and trends. *Resources and Environment, 03*, 48–55.

Teo, H. C., Lechner, A. M., Walton, G. W., Chan, F. K. S., et al. (2019). Environmental impacts of infrastructure development under the belt and road initiative. *Environments, 6*(6), 72.

The United Nations Framework Convention on Climate Change (UNFCCC). (2018). The People's Republic of China Second biennial update report on climate change. Retrieved 21 January 2021 from https://unfccc.int/sites/default/files/resource/China%202BUR_English.pdf.

United Nations (2015). Transforming our World: 2030 Agenda for sustainable development, Report Number A/RES/70/1. Available from: https://stg-wedocs.unep.org/bitstream/handle/20.500.11822/11125/unepswiosm1inf7sdg.pdf?sequence=1.

von Stechow, C., Minx, J. C., Riahi, K., Jewell, J., McCollum, D. L., Callaghan, M. W., Bertram, C., Luderer, G., & Baiocchi, G. (2016). 2 °C and SDGs: United they stand, divided they fall? *Environmental Research Letters, 11*(3), 034022.

Xie, H., Wen, J., & Choi, Y. (2021). How the SDGs are implemented in China: A comparative study based on the perspective of policy instruments. *Journal of Cleaner Production*, 125937.

XinHuaNet. (2015). China is taking action to tackle global climate change. Retrieved 19 February 2021 from http://www.xinhuanet.com/world/2015-11/28/c_128477994.htm.

XinHuaNet. (2020). Zhangjiakou has become the second wind power city in China whose installed capacity has exceeded 10 million kilowatts. Retrieved 19 February 2021 from http://m.xinhuanet.com/he/2020-06/07/c_1126083864.htm.

Xue, L., Weng, L., & Yu, H. (2018). Addressing policy challenges in implementing sustainable development goals through an adaptive governance approach: A view from transitional China. *Sustainable Development, 26*(2), 150–158.

Yang, X., Pang, J., Teng, F., Gong, R., & Springer, C. (2021). The environmental co-benefit and economic impact of China's low-carbon pathways: Evidence from linking bottom-up and top-down models. *Renewable and Sustainable Energy Reviews, 136*, 110438.

Yue, L. (2012). Dynamics of clean coal-fired power generation development in China. *Energy Policy, 51*, 138–142.

Zhang, Z. (2010). China in the transition to a low-carbon economy. *Energy Policy, 38*(11), 6638–6653.

Zhang, K. M., & Wen, Z. G. (2008). Review and challenges of policies of environmental protection and sustainable development in China. *Journal of Environmental Management, 88*(4), 1249–1261.

Zhang, L., Huang, Y. X., Li, Y. M., & Chen, X. L. (2010). An investigation on spatial changing pattern of CO_2 emissions in China. *Resources Science, 02*, 211–217.

Zhang, N., Lior, N., & Jin, H. (2011). The energy situation and its sustainable development strategy in China. *Energy, 36*(6), 3639–3649.

Zhang, S., Gao, Y., Feng, Z., & Sun, W. (2015). PPP application in infrastructure development in China: Institutional analysis and implications. *International Journal of Project Management, 33*(3), 497–509.

Zhang, Q., Liu, S., Wang, T., Dai, X., Baninla, Y., Nakatani, J., & Moriguchi, Y. (2019). Urbanization impacts on greenhouse gas (GHG) emissions of the water infrastructure in China: Trade-offs among Sustainable Development Goals (SDGs). *Journal of Cleaner Production, 232*, 474–486.

Zhang, C., Cai, W., Liu, Z., Wei, Y.-M., Guan, D., Li, Z., Yan, J., & Gong, P. (2020). Five tips for China to realize its co-targets of climate mitigation and Sustainable Development Goals (SDGs). *Geography and Sustainability, 1*(3), 245–249.

Zhangjiakou News. (2020). Great Wall International Renewable Energy Forum (2020) was held in Zhangjiakou, Hebei Province. Available from: https://unfccc.int/event/climate-action-summit-2019.

Zhong, Y. Y. (2016). Scenario analysis and path selection of China regional carbon emission based on low carbon development. *Ecological Economy, 32*(011), 71–74.

Zhou, X., Zhang, M., Zhou, M., & Zhou, M. (2017). A comparative study on decoupling relationship and influence factors between China's regional economic development and industrial energy–related carbon emissions. *Journal of Cleaner Production, 142*, 783–800.

Zhu, J. (2017). The 2030 agenda for sustainable development and China's implementation. *Chinese Journal of Population Resources and Environment, 15*(2), 142–146.

Zhuang, G. Y. (2007). Climate change challenge and China's low carbon economic development. *International Economic Review, 000*(005), 50–52.

Zhu, J., Sun, X., He, Z., & Zhao, M. (2019). Are SDGs suitable for China's sustainable development assessment? An application and amendment of the SDGs indicators in China. *Chinese Journal of Population Resources and Environment, 17*(1), 25–38.

Chapter 6
Urban Evolution Under Low Carbon Strategies

6.1 The Inevitable Low Carbon Future

In the Post-Paris Era, the participation of cities in climate governance has become one of the trending topics in tackling climate change (Cheshmehzangi, 2016). This participation is because of the natural driving force and executive power of coping with climate change and its impacts on living habitats (Wu & Zhu, 2020). In the context of moving towards a low carbon future, cities play a number of important roles. For instance, cities are pioneers of climate policy and technological innovation, laboratories of the urbanisation process, institutions conducting carbon-related experiments, grassroots mobilisation hubs, and civil society experimentation centers (Cheshmehzangi et al., 2018; Raven et al., 2019; Romero-Lankao et al. 2014). The concept of low carbon city, which is arguably mentioned concerning projects or researches of the World Wide Fund (WWF & CSUS, 2021), refers to the implementation of a low carbon economy within the city and keeping energy consumption and carbon dioxide emission at a low or even zero level. Therefore, it is important to see how cities and urban areas consider low carbon strategies in their steps towards achieving low carbon and zero-carbon targets.

To follow up on our earlier chapters, this chapter focuses on cities' irreplaceable role in a low carbon future. From the source of low carbon concept development to the phased evolution of low carbon cities, we delve into specific discussions on cities' position in low carbon and carbon neutrality directions. In doing so, we first elaborate on the concept and character of low carbon cities and shed light on some low carbon cities' evaluation methods. Later, this chapter spills a lot of ink over low carbon cities' main strategies and tools and illustrates the importance of low carbon concept planning and layout in cities by describing typical cases of global low carbon cities. We look into some historical progress in this area and highlight some key factors and strategies for low carbon city development. We discuss some international examples before exploring some adaptation strategies and low carbon initiatives. Lastly, we conclude this chapter with brief comparisons and summarising issues and prospects in achieving sustainability transitions.

A. Cheshmehzangi and H. Chen, *China's Sustainability Transitions*,
https://doi.org/10.1007/978-981-16-2621-0_6

6.1.1 The Concept and Character of Low Carbon City

Some scholars believe that a low carbon city is a sustainable ecological system that includes low carbon production and consumption (Fu, 2010). Other scholars believe that a low carbon city is a city dominated by a low carbon industry. Citizens take low carbon life as the concept and action guide, and the government takes low carbon society as the blueprint (Cheshmehzangi, 2020; Gu et al., 2009). A growing scientific consensus believes that a 'low carbon city' is to develop a low carbon economy with urban space as the carrier (Xia, 2008). In this regard, we can argue that low carbon urban development aims to achieve sustainable urban development by reducing energy consumption, pollutant emission, and greenhouse gas emission through the transformation of economic development mode, consumption concept, and lifestyle on the premise of ensuring the continuous improvement of urban residents' quality of life (Cheshmehzangi et al., 2021; Xin & Zhang, 2008). To be specific, low carbon cities include the following contents.

The first is to optimise industrial structure and develop low carbon industries. Secondly, the construction of low carbon infrastructure and buildings needs to be promoted. After that, reducing the use of high-carbon energy and increasing renewable energy consumption are both significant. Some policies should also cultivate the concept of low carbon life and advocate a low carbon lifestyle. Finally, energy-saving technologies, carbon capture, and storage technologies are adopted to increase the comprehensive utilisation of resources and pollution control (as shown in Fig. 6.1). Therefore, a low carbon city should have the following characteristics: (1) Economy (Efficient and intensive), (2) Security (Ecological, economic and social security), (3) Systemic (Complex systems involving different aspects of the economy, society, and environment), (4) Dynamic (Adaptability to dynamic change), and (5) Regional

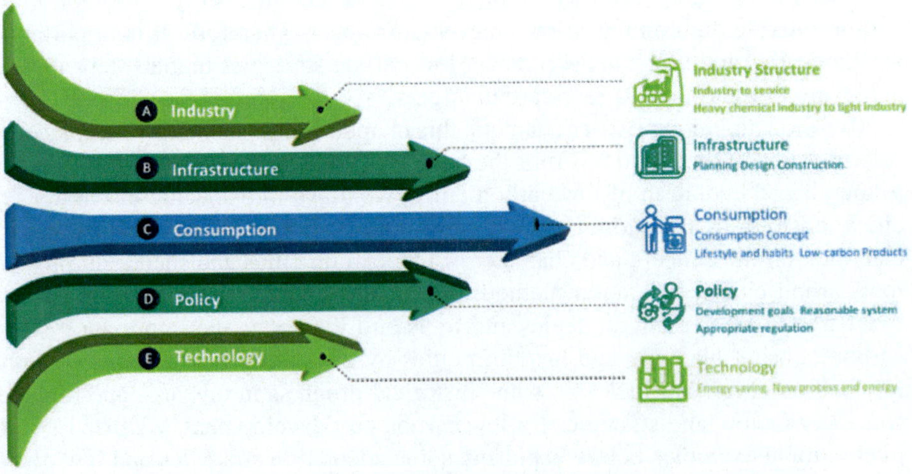

Fig. 6.1 Support system for low carbon cities (Drawn by the Authors)

(Urban and rural integration, coordinated development among cities of different scales). This leads to the follow-up discussion on evaluating low carbon cities, their strategies, and their impacts on achieving sustainable urban development.

6.1.2 Evaluation Method of Low Carbon City

At the early stage, the international evaluation of low carbon cities mainly adopted the Main Index Method (MIM). In other words, some individual indicators with the highest recognition and convenient statistics are used to describe the low carbon level achieved by the city. Common indicators include per capita carbon emissions, per unit domain area carbon emissions, per unit GDP carbon emissions, and local human development index (HDI) (Long et al., 2010). The main advantage of this method is that it can accurately reflect the city's low carbon level and the dynamic change of carbon emissions. The disadvantage is that there is no unified carbon emission coefficient or no spatial comparability, which will lead to a big difference in the low carbon level of the same city measured by different indicators. More scholars gradually recognised the Synthetic Index Method and Multi-Criteria Evaluation in recent years because evaluating the low carbon city needs both comprehensive measure and concrete analyses (Su et al., 2013). In other words, a variety of indicators or criteria related to urban low carbon development are selected for comprehensive analysis. For major greenhouse gases, the influencing factors mainly include population, energy consumption intensity, energy structure, or carbon content per unit of energy.

Moreover, there are many factors affecting urban carbon emissions, including consumption, construction, transportation, technology, and other fields. Therefore, the 'Synthetic Index Method' and 'Multi-Criteria Evaluation' have the advantages of extensive and comprehensive consideration, but they also have some defects. For example, it can only reflect the relative level of urban low carbon. For example, social, industrial, economic, and environmental performances are the target layers. The criterion layer refers to the influence of subsystems on various aspects of comprehensive evaluation indexes. Then the 'Entropy Weight Method' and 'Fuzzy Comprehensive Evaluation' gradually emerged (Azizalrahman & Hasyimi, 2018). The weight of each index layer can be determined comprehensively according to the Delphi method, and the comprehensive value of low carbon economy can be obtained by the linear weighting method (Ma et al., 2010). Although these evaluation schemes all have a common goal, that is, to measure the low carbon dimensions of various cities or to judge whether they meet the low carbon requirements, they have some differences in concept definition, indicator selection, methods, and operation modes (Table 6.1) (Tan et al., 2015). In sum, we verify the two assessment systems focus on low carbon cities at different stages.

Table 6.1 Comparison of different evaluation systems

Concern	Criteria or indicators	System construction method
Evaluation of low carbon level	Energy, transportation, science and technology, environment, economy, consumption	Analytic hierarchy process (AHP) Entropy weight method Fuzzy comprehensive evaluation
	Economy, industry, science and technology, society, environment	
	Economic development and social progress, energy structure and usage efficiency, Living consumption, development surroundings	
	Economic, energy, urban accessibility, carbon and environment, Technology development, waste, social and living	
Whether a particular city is low carbon or not	Economic, energy, land use, carbon and environment, transportation, waste, water	Multi-criteria evaluation

6.2 Histories and Typologies of Low Carbon Development

Cities are not only the main source of global GHG emissions but also the key to mitigating climate change. Many cities around the world have signed the Kyoto Protocol and made low carbon commitments (Chavez & Ramaswami, 2011; Cheshmehzangi & Dawodu, 2019). In this period, decoupling greenhouse gas emissions from urbanisation is a challenge, as all regions of the world increasingly confront the potential impacts of climate change (Chester et al., 2014). The interaction between urban and non-urban environments creates a variety of complex issues, including rising resource consumption, vulnerability to climate change, environmental pollution, and social inequality. In this context, the idea of low carbon transformation means that cities will move to a new, decarbonised path. Decarbonisation at the city scale requires a combination of technology development, infrastructure investment, and behavioural change (see Chap. 12 for further details). Thus, the low carbon transition strategies are necessary as part of the solution to achieve sustainable urban development (Olazabal et al., 2015). Improving the urban environment helps reduce greenhouse gas emissions. Measures to mitigate climate change in the built environment, the transport sector, power generation, and other areas can have a noticeable impact on the health of urban dwellers (Milner et al., 2012). Common and effective ways to achieve low carbon development include—but are not limited to—strengthening policy and regulatory frameworks, human capital development, investment in green technologies, and financial instruments (Lee, 2017). Urban transformation is used to effectively meet low carbon targets and align cities with sustainable development. This transformation is tangled with the idea of low carbon city progression.

6.2.1 *Rise and Evolution of Low Carbon Cities*

Two important international conventions, the United Nations Framework Convention on Climate Change (UNFCCC) in 1992 and the Kyoto Protocol in 1997, foreshadowed the origins of low carbon development (Zeng & Zhang, 2011). Since the UK first put forward the concept of a "low carbon economy" in 2003, the concept of low carbon has gradually gained popularity (The UK, 2003). The research on the connotation of low carbon city has also evolved into a comprehensive topic covering society, culture, economy, and environment instead of an initial environmental topic focusing on reducing carbon emissions. After the G8 Summit in Italy in July 2009, which proposed the goal of a 50% reduction in GHG emissions by 2050, climate change became a hot topic in addition to economic issues (Fang et al., 2009). The leaders of 16 countries, including Australia, Brazil, China, the United Kingdom, and the United States, all announced that they would implement low carbon growth plans. At the Copenhagen Conference in December 2009, world leaders proposed emission reduction targets (Dimitrov, 2010), which also faced some disagreements between developed and developing nations. In recent years, global environmental organisations such as the United Nations Environment Program (UNEP), the European Environment Agency (EEA), and WWF for Nature have become more concerned about the impact of climate change on living habitats, including the human habitats. They believe that energy conservation, emission reduction, and building low carbon cities are important measures to mitigate the current global warming crisis. In particular, UNEP (2013) refers to low carbon as in the following:

> The term 'low carbon' has also changed with the growing evolving understanding and assessment of 'climate sensitivity'. Typically, 'low carbon' transformation refers to global GHG emissions pathways that stabilize GHG concentration in the atmosphere at levels considered not to be dangerous anthropogenic interference with the climate system.

At the same time, research and practice on the low carbon economy and low carbon city are also emerging fast in many countries. For example, the Theme of World Environment Day 2008 was determined by the United Nations Environment Program as "Change the traditional concept and promote low carbon economy". The ISOCARP held its 46th annual meeting under the theme of "Low carbon Cities" in October 2009 in Porto, Portugal (Yokota, 2010). After that, low carbon cities began to be further promoted and expanded in the world. In facing the climate change impacts on cities, the position of low carbon urbanism or low carbon city development has become more evident in the field of sustainable development.

6.2.2 *International Low Carbon City Development Model*

The authors conclude that the development types of low carbon cities can be divided into two categories as follows.

1. **By City Scale (According to population, land use)**
There is a strong correlation between emissions and land use regulations or population. New, old, and suburban areas all have different characteristics (Glaeser, 2010). The first is the densely populated megacities or clusters of cities dominated by a few large cities. This kind of centralised urbanisation development has obvious advantages in improving economic efficiency and energy efficiency, saving resources, protecting the environment, and controlling the loss of cultivated land, which are of great benefit to low carbon development. Besides, the high population density and high floor area ratio (FAR) of compact urban form can bring many benefits to the city (Butters et al., 2020; Dawodu & Cheshmehzangi, 2017; Wu, 2014). On the environmental front, it will shorten the distance between cities, improve transportation accessibility, reduce car dependence and energy consumption, and help protect cities around arable land and maintain biodiversity (Cheshmehzangi & Butters, 2017). Besides, improved efficiency in the use of infrastructure and reduced maintenance costs make it easier for urban residents to access a variety of local jobs and public services. The development of other types of low carbon cities is more decentralised. For example, distributed growth in American cities or smaller cities in Germany.
2. **By Level of Development (Categories like developed, developing, underdeveloped)**
Due to the difference in national income, the development stages of low carbon cities can be divided into developed, developing, and underdeveloped economies. Economically developed cities tend to take the lead in making low carbon plans, and high-income countries are relatively early in making carbon reduction commitments. In some developed countries, low carbon cities, low energy consumption, high value-added industries or services have replaced traditional industries, like Malmö in Sweden. Simultaneously, low carbon cities in some high-income countries only have service industries, which are typical zero-carbon cities, such as Masdar City in the United Arab Emirates (Long et al., 2010). Nevertheless, the progress of zero-carbon cities such as Masdar City is not the same as what has been proposed initially. Moreover, in some low carbon cities in developing countries, their forests, farmlands, and other ecological resources are well preserved, and their total carbon emissions are low.

6.3 Key Factors and Strategies

The development strategies of international low carbon cities vary, but some common features can be found in some successful cases. Firstly, low carbon city construction has a clear goal, especially the quantitative indicators of emission reduction (Cheshmehzangi et al., 2018). This has pointed out the direction of efforts for the effective development of low carbon city construction (Qiu & Li, 2011). The second focuses on production, life, transportation, other economic and social life, and other fields. Decarbonisation strategies can be implemented in terms of population density, climate, energy mix, and transportation and building-related energy consumption patterns (Kennedy et al., 2014). In addition, a variety of policy tools are used comprehensively in the construction of low carbon cities. System and policy design can mobilise the enthusiasm of relevant government agencies, enterprises, and the public. Finally, the system, regulations, and documents play a key role in supervising the implementation (Chen, 2011). Romero-Lankao (2014) proposed that the mechanism of cities and carbon cycle could be summarised as shown in the following:

- Multiscale interactions among social, ecological, and technological systems;
- Mechanisms by which the exchange of carbon flows inside and outside cities affects atmospheric, terrestrial and aquatic systems;
- Carbon feedbacks affect urbanisation;
- Interactions between urbanisation, urban regions, and carbon.

The carbon constraint of urban development caused by global climate change is gradually emerging. More and more cities around the world have actively joined the ranks of addressing climate change and gradually implemented low carbon transformation in terms of target setting, planning, and implementation of measures (WWF, 2020). Low carbon cities are spreading across the globe. Many cities first put forward low carbon construction targets and then made related plans or action plans. Therefore, low carbon city strategies often focus on climate, environment, carbon use, society, technology, and economy (Evans et al., 2014). The first wave of cities to carry out low carbon planning globally are mainly the members of 'Large Cities Climate Leadership Group (C40)'. These Cities have entered the implementation stage of low carbon city construction goals, including London, New York, Copenhagen, Tokyo, Toronto, Amsterdam, Austin, Stockholm, etc. Most of these cities entered the post-industrial stage before they set the target of low carbon and were already ahead of the world to renew energy sources and protect the environment. Therefore, we can see an obvious difference between cities of developed and developing countries.

6.3.1 Governance: The Power of the Policy

Most of the low carbon cities in the world have a similar process of policymaking. The first step is to understand a city's carbon footprint and establish a framework and benchmark for accounting for carbon emissions. Then it sets low carbon targets and analyses the results under different scenarios (Cheshmehzangi, 2020; Harris et al., 2020). The penultimate step is to develop action plans for cities to reduce emissions. Finally, yet importantly, to ensure the implementation of the plan, mechanisms are needed to monitor and assess emissions (Hasyimi & Azizalrahman, 2018). In this regard, appropriate policy is the prerequisite and basic tool to achieve the emission reduction target. The authors collected the low carbon policies of some countries, as shown in Table 3.2.

In terms of implementation effect and final effect, mitigation and adaptation strategies can be divided into different categories. These categories include specific areas of policy functions, including land use, transportation, construction, natural resources, renewable energy sources, water supply, and waste management. More importantly, low carbon policies have corresponding specific measures and action contents and pay attention to the synergistic effect between different fields' measures. This is commonly practiced to maximise the economic and social benefits of low carbon development. For example, the UK's WWF climate and energy team focuses on three main areas of power, transport, and buildings, which account for 84% of the UK carbon emissions (WWF, 2021). To expand on such country-level examples, the authors collected some important low carbon policies of different countries in the last 10 years. From the findings, we can see that countries at different stages of development have different priorities. These could be due to different proportions of their industries or the richness of their resources. But, they also have a lot in common, especially in the phase change of low carbon policy. Lastly, many countries are increasingly focusing on the management of enterprises and the guidance of individuals. The findings are summarised in Table 6.2.

6.3.2 Planning and Low Carbon Targets

Low carbon urban planning is an important regulatory means for the government to guide urban development. By utilising the city's rational layout, various functional systems of the city can run well and reduce carbon emission in the process. The extensive production mode of high energy consumption, high pollution, and low efficiency has been abandoned. The development mode of low input, high output, high benefits, and low pollution has been transformed (Price et al., 2013). Main planning strategies include those criteria shown in Fig. 6.2, namely energy, finance, buildings, traffic, citizens, and waste. For example, low carbon environmental protection products should be promoted according to residents' low carbon concepts and lifestyles. New efficient and clean energy can be discovered, or the cost of using new energy can be reduced. In achieving low carbon targets, buildings and transportation should maintain low energy consumption and low carbon emissions (Ho et al., 2013). In this process, we expect governments to improve the efficiency of traditional energy systems in many ways and guide the development of public transport.

Taking some typical successful low carbon cities as examples, we can conclude that low carbon strategies play an irreplaceable role in guiding cities' sustainable development. The following ten pioneering city examples represent different models of successful low carbon development. We can also learn from such examples through further evaluation of their strategies, priorities, and practices.

Amsterdam, the Netherlands

Amsterdam aims to meet the World Health Organization (WHO) air quality guidelines of 2030 and plans to reduce carbon dioxide emissions by 9%. Amsterdam is

Table 6.2 Low carbon policies in selected countries or regions from 2010 to early 2021 (Collected by the Authors using country-level data extracted from https://www.iea.org/)

Country or region	National policies (Some of important or still in force)	Year
United States	The utilising significant emissions with innovative technologies (USE IT) Act	2019
	Updates to the minimum cooling efficiency standards for air conditioning and heating equipment	2018
	Alternative fuel vehicle community partner projects	2017
	Greenhouse gas emissions and fuel efficiency standards for medium- and heavy-duty engines and vehicles—phase 2; solar energy technologies office/sunshot; water power technologies office (WPTO)	2016
	Renew300 federal renewable energy target	2015
	Rural energy savings program	2014
	Energy Efficiency and conservation Loan Program; US Climate Action Plan	2013
	High energy cost grant program; U.S. Africa Clean Energy Finance (US-ACEF) Initiative	2012
	Better buildings initiative	2011
	Mandatory reporting of greenhouse gases	2010
United Kingdom	Carbon capture and storage support in the 2020 UK budget	2020
	Electric vehicle homecharge scheme	2019
	UK CCUS Action Plan	2018
	Clean growth strategy and industrial strategy; UK clean growth strategy; ultra low emission trucks scheme; CO_2 emission based vehicle tax rates from 1 April 2017	2017
	Accelerating carbon capture and storage (CCS) technologies (ACT) Electric vehicle homecharge scheme (EVHS) and workplace charging scheme (WCS) chargepoint authorisation	2016
	Long-term investment in ultra-low emission vehicles in the UK	2015
	Climate change agreements Energy saving opportunities scheme (ESOS); low carbon vehicle innovation platform (LCVIP); ultra low emissions vehicles (ULEV) programme; UK National Energy Efficiency action plan 2014	2014
	2nd national energy efficiency action plan; energy white paper 2011; national energy efficiency database	2011
	Carbon reduction commitment energy efficiency scheme (CRC); national renewable energy action plan (NREAP)	2010
Germany	Climate protection projects in social, cultural and public institutions; Industrial incentive program to increase energy efficiency and the use of renewable process heat	2019
	Electric vehicle support for public transport; the electricity grid action plan	2018
	2017 amendment of the renewable energy sources Act (EEG 2017) (Federal Ministry for Economic Affairs and Energy, 2017)	2017

(continued)

Table 6.2 (continued)

Country or region	National policies (Some of important or still in force)	Year
	Energy efficiency incentive programme; promotion of exemplar climate mitigation projects by municipalities; model climate protection; projects in communities; heating system optimisation	2016
	Energy Efficiency networks initiative	2015
	3rd national energy efficiency action plan (NEEAP); energy conservation regulations (EnEV) 2014	2014
	Energy and electricity tax cap; support of energy-efficient and climate-friendly production processes; support of energy management systems	2013
	Electricity saving initiative	2012
	Law on energy and climate fund	2011
	Energy and climate act, 2010; national energy action plan (NREAP)	2010
Denmark	Danish climate agreement for energy and industry etc. 2020 of 22 June 2020 (only EE dimension); energy savings in state owned institutions; Promoting energy efficient refurbishments (campaign); Building hub	2020
	Danish energy agreement of 29 June 2018 (only EE dimension)	2018
	Green owner tax; Purchase tax rebate vehicle Denmark; reduction in vehicle registration tax for EVs	2017
	Danish building regulations 2015 (BR15); voluntary energy saving agreement: electricity-intensive companies	2015
	Danish building regulations br10; energy strategy 2050	2011
	Danish energy saving trust; the energy saving council; green growth agreement 2010–2012	2010
Sweden	Environmental zones for municipalities	2020
	Ordinance (2019:525) on state aid for the installation of charging points for electric vehicles	2019
	Bonus-malus system for passenger cars, light trucks and light buses; Draft Sweden national biogas strategy; ordinance (2015: 579) on support for promoting sustainable urban environments; The Swedish Climate Act	2018
	Coaches for climate and energy; law on energy mapping in large companies; support for retrofits and energy efficiency in certain areas	2016
	Drive Sweden; Klimatklivet; Local Climate Investment programme (Climate Leap)	2015
	Super green car circulation tax; super green car rebate; voluntary certification of installers of certain heating systems	2012
	The super-green car premium; the fuels act; infrastructure investment; exhaust emission controls act; environmental requirements in the procurement of vehicles	2011
	Environmental vehicle premium; energy audits for companies	2010

(continued)

Table 6.2 (continued)

Country or region	National policies (Some of important or still in force)	Year
Japan	Japan's long-term strategy under the Paris Agreement (Cabinet decision, June 11, 2019); roadmap for Carbon Recycling technologies; strategy for developing hydrogen and fuel-cell technologies; roadmap for carbon recycling technologies	2019
	Long-term goal and strategy of Japan's automotive industry for tackling global climate change; strategic energy plan (2018); act on the rational use of energy (revision of Energy Conservation Act); Tax system for promoting energy efficiency	2018
	The plan for global warming Countermeasures; subsidies for commercial and residential building energy efficiency investments; labelling system for energy efficiency; building energy efficiency act	2016
	Long-term energy supply and demand outlook; subsidies for new clean energy vehicles	2015
	Basic energy plan (2014); strategic energy plan (2014)	2014
	Development of innovative energy efficiency technologies; feed-in tariff for renewable electricity and solar PV auction; innovative strategy for energy and the environment; low carbon city act (Eco-City Act)	2012
	National electricity saving action	2011
	Promotion of zero energy building (ZEB) and zero energy houses (ZEH); standards of judgment for construction of specified buildings (2010)	2010
Switzerland	Switzerland Energy act of 30 september 2016; law on energy; market premium for large-scale hydropower	2018
	Energy Strategy 2050; CO_2 emission standards for light duty vehicles	2017
	Action plan for a coordinated swiss energy research; technology fund for innovative technologies; obligation for CO_2 compensation by fuel importers	2013
	CO_2 emission standards for passenger cars	2012
	Building refurbishment program; tenders for efficient use of electricity	2010
South Africa	AFD green fund; integrated resource plan 2019 (IRP 2019) (Department of Mineral Resources and Energy, 2019); South African Carbon Tax; South Africa's low-emission development strategy (SA LEDS) 2050	2019
	Compulsory specifications for energy efficiency and labelling of electrical and electronic apparatus (VC 9008)	2016
	National Energy Efficiency strategy post 2015	2015
	South African energy Efficiency Label—electric ovens	2014
	Global climate partnership fund	2013

(continued)

Table 6.2 (continued)

Country or region	National policies (Some of important or still in force)	Year
	SANS 204: energy efficiency in buildings; energy efficiency management; renewable energy independent power producer programme (REIPPP);	2011
	Renewable energy and energy efficiency partnership; industrial energy efficiency programme; GEF special climate change fund	2010
Netherlands	Environment and planning act; industry carbon tax; National CO_2 pricing system for industry	2021
	HER+ Subsidy scheme renewable energy transition; Minimum CO_2 price electricity production; Prohibiting/closing down coal fired power plants; sustainable energy transition scheme (SDE++); Stimulation scheme natural gas free rental housing	2020
	Climate act; climate agreement; duty to save energy—communication campaign; DEI+ Demonstration scheme energy and climate innovations; national agenda on charging infrastructure; Subsidy scheme reduction energy use;; subsidy scheme for long term mission oriented innovation program for built environment; subsidy scheme on energy efficiency and renewable energy in horticulture (EHG)	2019
	Energy label targets; government-wide programme for a circular economy; green deal 221 participation of communities in sustainable energy projects; programme and large scale pilots on natural-gas-free neighbourhoods	2018
	Demonstration scheme climate technologies and innovations in transport; electricity tax breaks for providing public EV charging infrastructure	2017
	Administrative agreement on zero emission buses; energy-saving at home subsidy scheme (SEEH); green deal on electric transport 2016–2020 (Green Deal 198); green deal on zero emission city logistics (Green Deal ZES); Long-term industry agreements on energy efficiency (LTA3 / MEE)	2016
	Company tax benefits for zero-emissions vehicles; green deal on public charging infrastructure; Offshore wind tender regulation; Netherlands offshore wind energy act (Wet Wind op Zee)	2015
	Energy audits EED; energy-saving fund for loans in the rented sector (FEH); national energy saving fund; The Netherlands's Energy consumption target	2014
	Agreement on energy for sustainable growth; energy agreement for sustainable growth; national energy agreement for sustainable growth	2013
	Energy Performance standards for buildings (EPG); technical building regulations from the point of view of energy efficiency and the environment	2012

(continued)

Table 6.2 (continued)

Country or region	National policies (Some of important or still in force)	Year
	EU CO_2 emission standards for light commercial vehicles (vans) (510/2011); green deals; offshore wind energy green deal; sectoral emission trading system in horticulture; support scheme for solar panels	2011
	International energy programme; national renewable energy action plan (NREAP); trucks for the future	2010
New Zealand	Clean car standard and clean car discount	2021
	State sector decarbonisation programme	2020
	EVRoam; warmer kiwi homes	2018
	Access to special vehicle lanes; improve programme operated by the energy efficiency and conservation authority (EECA); energy Levies—energy innovation (Electric Vehicles and other matters) Amendment Act 2017; low emission vehicles contestable fund	2017
	Warm up New Zealand: healthy homes extension for rentals programme; electric vehicles programme New Zealand	2016
	Carbon reduction programmes; air conditioners and chillers and air conditioners and chillers—updated policy proposals	2014
	Warm up New Zealand: healthy homes	2013
	New Zealand Energy Strategy (NZES); New Zealand energy efficiency and conservation strategy; national policy statement for renewable electricity generation	2012
	Emissions trading scheme; electricity industry act 2010	2010
Finland	Phasing out coal in energy production; promoting the use of biofuels in the transport sector	2019
	Nearly zero-energy regulations; promoting renewable energy (electricity)	2018
	Energy efficiency agreements, term 2017–2025 (Voluntary energy efficiency agreements); government report on medium-term climate change plan for 2030—towards climate-smart day-to-day living	2017
	National energy and climate strategy of Finland for 2030	2016
	Finland tender-based feed-in premium scheme for renewable power generation (30.12.2010/1396); EVELINA—national test environment for electric vehicles	2011
	A group of energy efficiency measures in agriculture	2010

already a leader in electric vehicles (EVs) and will ramp up its efforts in the coming years. Electric car drivers will have priority parking and loading space. Amsterdam currently has six low-emission zones, which ban the most polluting vehicles. They will gradually become zero-emission zones, where no use of petrol, diesel, or petrol vehicles will be allowed (The City of Amsterdam, 2021). From 2030, the entire built-up area of Amsterdam will have zero emissions for all vehicles, including cars and motorcycles.

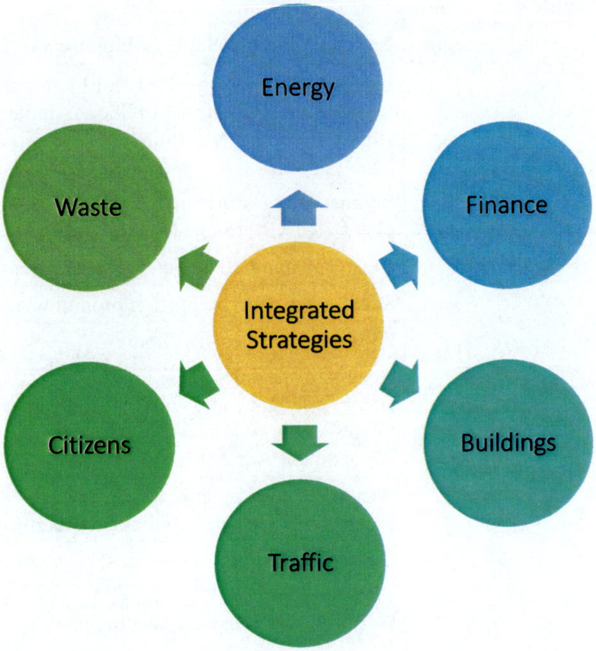

Fig. 6.2 Integrated strategies for low carbon cities (Drawn by the Authors)

Berlin, Germany

By 2050, Berlin plans to cut its carbon emissions by 85% of the 1990 levels. In October 2017, Berlin became the first German federal state to pass a coal abolition law, ending coal-fired power generation and heating by 2030. Distributed gas-fired cogeneration plants will be built to replace coal-fired plants that have been shutting down to generate electricity and provide district heating in the city. Berlin has launched extensive education and communication strategies among the public and businesses. It has launched the 'Save the Climate Book' and Berliners can use 'Green Rewards' cards to collect points for green spending and then cash them out. Such initiatives help mainstream 'Climate Neutral' as a common concept, which could also promote the message to different sectors of society, such as the 'Berlin Energy Efficiency Campaign'. This campaign is one of the projects aimed at the general public, which is aimed to provide advice and network development to small and medium-sized enterprises (SMEs) to promote climate-friendly innovation projects.

Copenhagen, Denmark

Copenhagen undertook the responsibility of addressing climate change through the CPH 2025 Climate Plan. The CPH 2025 Climate Plan includes specific targets and initiatives in four areas: energy consumption, energy production, green transportation, and city management (State of Green, 2021). Thanks to the rapid development of wind power installations in 2011, Copenhagen has already achieved ahead of schedule its target of reducing carbon emissions by 20% by 2015. There are now wind turbines and solar panels on the roofs of apartment buildings and villas all over Copenhagen. Copenhagen promotes non-motor transport, with nearly a third of its people commuting by bike. In addition, the development of energy-efficient buildings, the development of new energy sources, the increase of garbage recycling, and the expansion of forest area are also part of the city's energy conservation and emission reduction measures.

Hong Kong SAR, China

In 2017, the Hong Kong Government released the Hong Kong Climate Action Blueprint 2030+, aiming to reduce carbon intensity by 65–70% from 2005 levels in 2030. This is equivalent to an absolute carbon reduction of 26–36%. This is planned in the city's climate action plan, while per capita carbon emissions will be reduced from about 5.7 tons in 2015 to between 3.3 and 3.8 tons in 2030 (HKEB, 2017). At the moment, about 67% of Hong Kong's carbon emissions come from electricity generation, 90% of which is consumed by buildings (HKCSD, 2019). This is the reason why a primary focus is given to the building design, renovation, and energy-efficient built environments.

London, UK

London, which is the first mega-city to commit to reaching the World Health Organization's air quality guidelines, has put forward a series of low carbon action plans to deal with global climate change. Some of its primary action plans include 'London Plan' in 2004, the 'Mayors Energy Strategy and Climate Change Action Plan-Action Today', and the 'Mayor's Climate Change Action Plan' in 2007. The city's 'Climate Change Mitigation and Energy Strategy' was released in 2010. Therefore, we can verify London has become a leader in improving the global environment. It puts forward the distributed energy supply strategy and the community energy conservation activities to support the creation of low carbon communities. In the area of architecture and built environment, the Green Institution Action Plan was adopted. In terms of transportation, London implemented measures to reduce motor vehicles' carbonisation and fuel and levy traffic congestion charges (UNEP, 2020). In this area, London has been partly successful and continues to improve its low carbon targets and integrated strategies.

New York, USA

In 2002, New York joined the International Council for Local Environment Initiatives (ICLEI) and launched the Cities for Climate Protection Campaign (CCP) (Rickerson & Hughes, 2006). In 2007, New York City released the first version of its climate plan, "PlaNYC—a Greener, Greater New York", which for the first time included reducing greenhouse gas (GHG) emissions as a goal of its commitment. 'One New York: The Plan for a Strong and Just City' (One NYC), which was released in 2019, proposes the city to be carbon neutral by 2050. Its key policies include reducing greenhouse gas emissions from buildings, using 100% clean electricity, creating green jobs, and holding polluters accountable. In 2016, the "New York City 80 by 50 Roadmap" was released, laying out a deep roadmap to reduce emissions.

Paris, France

In April 2009, France announced the new "Plan of Grand Paris" for the transformation of Paris. The "low carbon" concept of urban transformation provides a good reference for the future development of the metropolis. Some advanced ideas include limiting urban sprawl, changing traffic patterns, and leading low carbon lifestyles. In September 2018, 19 megacities, including Paris, signed a "Declaration on Net Zero Carbon Buildings," pledging to achieve net-zero carbon emissions for all buildings in the city by 2050 (Scott, 2018). The Paris Climate Action Plan sets zero-carbon targets for energy, buildings, urban planning, waste, mobility, food, and so on (City of Paris, 2018). This plan further clarifies the vision and steps of Paris to reduce carbon.

Seoul, South Korea

The city of Seoul, South Korea, launched its Nanum-Car program in 2013 as part of its transport demand management strategy. To encourage the use of environmentally friendly cars, Seoul has launched an electric-car sharing service. Seoul Metropolitan Government (2015) proposes plans covering energy, air quality, transport, waste recycling, water, ecology, urban architecture, health, and urban planning. South Korea has pledged to be a role model in tackling climate change as part of its preparations for a new post-2020 climate regime.

Singapore

The Singapore Sustainability Blueprint details more stringent environmental targets by 2030, doubling the amount of greenery in the city (through green façades, green roofs, roof gardens, etc.) to nearly 500 acres and adding more than 50 miles of travel along the green passage at the park connector. Singapore, which gets 95 percent of its energy from natural gas, has begun pushing a goal to increase solar panel coverage. In 2014, Singapore's Economic Development Board launched SolarNova, a project that invites bids from companies to install solar panels. Singapore has also decided to take big steps to improve energy efficiency. In doing so, the primary aim is to reduce energy intensity by 35% from 2009 levels by 2030. Singapore strictly controls the ownership and use of private cars to reduce carbon emissions from the transport sector. First, the vehicle quota and ownership license system. The second, the road pricing system. To date, Singapore remains a pioneering city in this area. In 2001, Singapore launched the green car tax rebate scheme to encourage the development of green cars.

Tokyo, Japan

As the birthplace of the Kyoto Protocol (i.e., Japan), Tokyo issued an environmental draft in 2002 and the Tokyo Policy on Climate Change in 2007. Tokyo caps emissions for industrial and commercial enterprises with high emissions, It also promotes a mandatory environmental mortgage insurance scheme for small and medium-sized enterprises (SMEs). At the same time, Tokyo has launched a series of measures to reduce household energy consumption and promote solar products into thousands of homes. Tokyo has also increased the proportion of energy-efficient buildings, set building energy efficiency technical targets, fuel efficiency standards for vehicles, and encouraged the use of environmentally friendly and green fuels (Fig. 6.3).

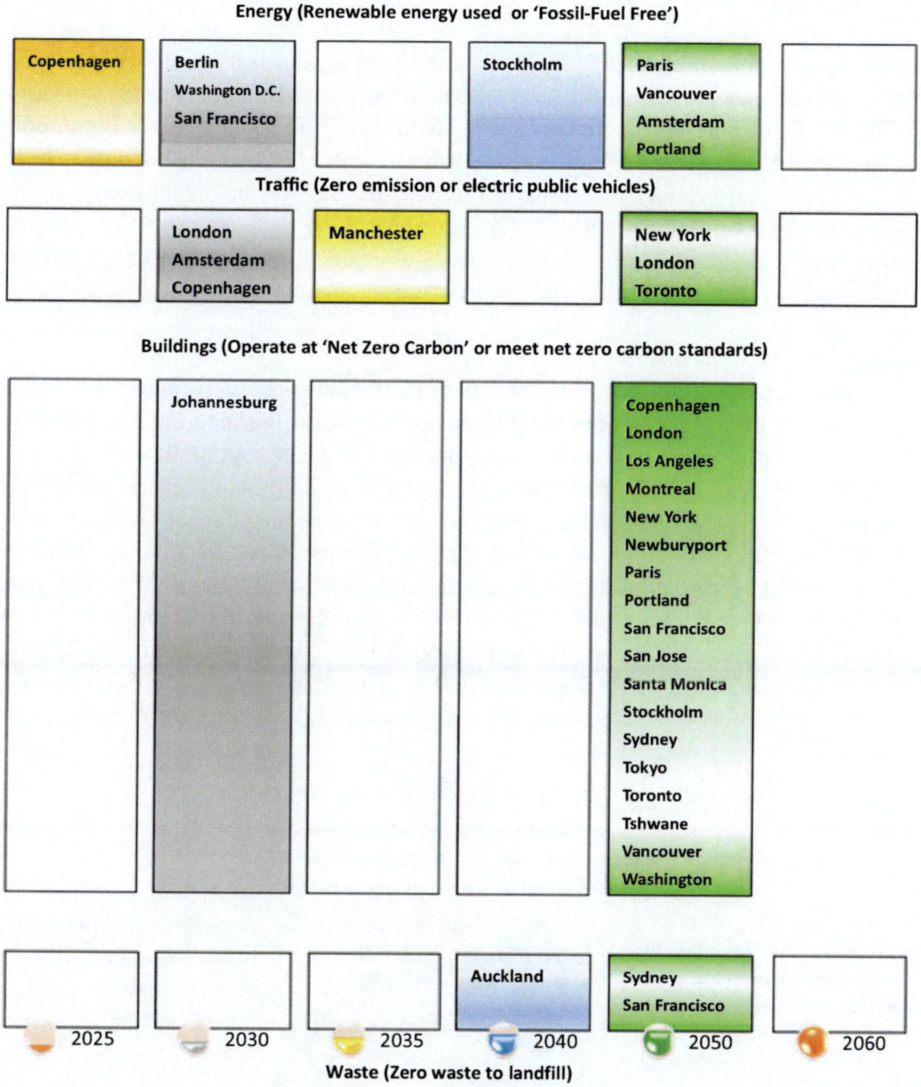

Fig. 6.3 Low carbon target years for selective global city examples (Drawn by the Authors)

6.4 Adaption to Low Carbon: Practices and Challenges

6.4.1 *Low Carbon Urban Initiatives*

Low carbon urban initiatives (LCUIs or initiatives) are defined as initiatives in cities that integrate climate mitigation strategies in urban development projects (van Doren et al., 2018). Important features of LCUIs are that they are initiated at the community scale rather than at the individual household level, which has benefits not only in terms of carbon reduction but also in terms of reducing transaction and instalment costs and strengthening community networks and ownership. In Fig. 6.4, we summarise the primary points of LCUIs, including publicity and education, buildings and facilities, energy and cooperation, traffic and transportation, and economy and financing.

Energy and Cooperation

Countries around the world have taken steps to reduce their reliance on coal. In 2017, for example, The United Kingdom and Canada co-founded the Powering Past Coal Alliance (PPCA), whose members pledged to phase out coal power by 2030 or 2050 (PPCA, 2021). Sweden closed its last coal-fired power plant in April 2020. Denmark claims to stop licenses for oil and gas exploration and fossil fuel production by 2050 (Climate Home News, 2020). Renewable energy has become an important choice for all countries to tackle climate change due to its wide distribution, great potential, and sustainable utilisation. For example, Germany has the largest scale of renewable energy development in Europe. In 2019, Germany issued the Climate Action Law and Climate Action Plan 2030, which clearly stated that the proportion of renewable energy generation in the total electricity consumption would increase

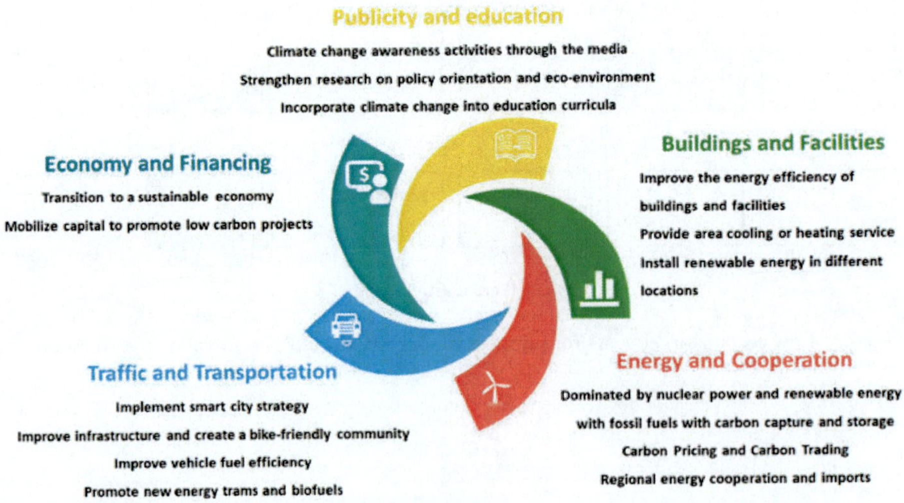

Fig. 6.4 The main points of LCUIs (Drawn by the Authors)

year by year. The proportion will reach over 80% by 2050 (CLEW, 2020). The United States enacted the "American Recovery and Reinvestment Act of 2009", focusing on encouraging private investment in wind power through tax credits and loan concessions (Congress, 2009). The EU released the Hydrogen Strategy in July 2020 to promote the development of hydrogen technology (EU, European Commission, 2020). Other countries like Denmark and the United Kingdom have proposed the development of hydrogen energy for industry, transportation, electricity, and residential power supply.

Buildings and Facilities

First, we introduce a green building evaluation system and promote green energy efficiency labeling. In terms of the evaluation system, the UK has introduced BREEAM, the first green building evaluation method in the world (BREEAM, 2021). Germany launched the second generation of green building evaluation system DGNB, covering ecological protection and economic value (DGNB, 2021). Singapore has included minimum Green standards in the Building Control Act, introduced the Green Mark evaluation system, and stipulated energy efficiency standards for new buildings, existing buildings, and communities (Singapore Statutes, 2019). In terms of green energy efficiency labeling, the United States and Germany have respectively implemented the "Energy Star" and "Building Energy Qualification Certificate" to mark the energy efficiency and consumable grade of buildings and equipment. In 2020, the European Commission launched the "Wave of Innovation" initiative, which proposed that all buildings would achieve near-zero energy consumption by 2030. France has set up renovation grants to help 7 million high-energy homes meet low-energy building standards.

Traffic and Transportation

Germany increased electric vehicle subsidies, Norway and Austria exempted zero-emission vehicles from value-added tax, the United States provided low-interest loans to vehicle companies that developed new technologies. For example, Costa Rica gave preferential tariff treatment and parking priority to citizens who bought zero-emission vehicles. Major developed countries and developing countries such as Mexico and India have announced a timetable for banning the sale of fuel vehicles. Low carbon fuel standards and tax credits in the United States. Japan, Chile, Peru, South Africa, Argentina, Costa Rica, and other governments issued green transportation strategies or traffic laws, unified car purchase standards, encourage the use of electric or zero-emission vehicles. Zero-emission vehicles are also being promoted in the field of land and water transport. The EU plans to invest €2.2 billion in 140 key transport projects through its "Connecting Europe Facilities (CEF)" (EU, 2021). Within Europe, digital technology will be used to build a unified ticketing system, expand the scope of traffic management systems, strengthen ship traffic monitoring and information systems, and improve energy efficiency.

Economy and Financing

European Commission issued Circular Economy Action Plan, which is through the product especially for electronics, batteries and automotive, packaging, plastics, and food. European Union recycling electronic plan, new battery regulatory framework, new mandatory requirements, and reduced packaging and plastic disposable packaging and containers aim to improve product circulation utilisation rate and reduce the European Union's carbon footprint (EERA, 2020). The UK also levies taxes on different sources of energy and sets up a carbon trust fund. Germany and Japan have similar low carbon fiscal and tax policies.

Publicity and Education
To promote energy conservation in various sectors, key stakeholders such as non-governmental organisations (NGOs) and educational institutions (such as universities) play their part effectively. The government encourages the whole society to participate in the energy-saving Target by setting targets and timelines and ensuring transparent disclosure of low carbon and energy-saving achievements (OECD, 2015). For example, "Low Carbon Scotland: A Behaviours Framework" (Scottish Government, 2010) popularised low carbon knowledge and habits to the public from the perspective of behaviour control and guidance.

6.4.2 Progresses and Comparison

The authors collected several low carbon strategies of selective city examples in different countries. This selection benefits from a comparative analysis and evaluating their achievements (or advantages), demonstration (or disadvantages), and specific characteristics. The comparison also includes policy types of these city examples in seven countries of Australia, the Netherlands, Denmark, Finland, France, Singapore, and Japan. Based on the comparison of those city examples, we could conclude they focus on similar areas or issues, such as renewable energies, green buildings, low carbon transportation, and sustainable industries. Besides, some cities propose specific strategies, such as green taxing systems or other financial policies and mechanisms. Some city examples also face similar problems, such as implementing climate change action plans for adaptation strategies (e.g., seen in Finland, Japan, and Singapore) and/or neglecting the significance of such plans in their further plans.

In sum, we note some general compatibilities and specific differences between different locales, suggesting a tangible divergence in pathways and directions in achieving the same or similar targets. Table 6.3 summarises these in seven city examples of Adelaide, Amsterdam, Copenhagen, Helsinki, Paris, Singapore, and Tokyo.

Table 6.3 Comparison of low carbon strategies in some countries or cities

Country	Policy types	Characteristics	Achievements or advantages	Demonstration or disadvantages
Adelaide (Australia)	Energy and buildings, transportation & urban planning, food waste and water, and carbon offsetting	Under the world's first carbon–neutral city goal, Adelaide will be a model for tackling climate change and seizing economic, social, and environmental opportunities. It will also develop renewable energy and clean smart technologies (Adelaide City Council, 2015)	• Encourage community action such as reward and support for individuals or researches • Energy Efficiency and renewables such as investment in solar PV, battery storage, energy efficient products, and electric vehicle recharging points • Investment in large scale renewables to reduce the emissions intensity of electricity supply; • Reduce emissions from waste to provide new recycling service and improve the use of low carbon materials; • Transform the travel to low emission public transport system and invest in cycling • Offset carbon emissions	The local government claims to be the first carbon–neutral city

(continued)

Table 6.3 (continued)

Country	Policy types	Characteristics	Achievements or advantages	Demonstration or disadvantages
Amsterdam (Netherlands)	Energy & Buildings, and Transportation & Urban planning	Amsterdam plans to eliminate all carbon dioxide emissions by switching from coal, oil, and gas to 100 percent clean energy. (Amsterdam City Council, 2019)	• Renewable energy. Gas will be phased out and geothermal heating is explored. About 17 MW of wind turbines will be installed by 2022. The city will realise the use of 250 MW solar energy • Green transportation. promote electric vehicles and public transport and give priority to the use of electric taxis and strengthen bicycle infrastructure • Sustainable industries. Dramatically reduce the use and transportation of fossil fuels by 2030. It finally aims to achieve sustainable ports by 2050. It will shut down fossil power plants and replace them with sustainable energy sources	A gas-free city by 2040

(continued)

Table 6.3 (continued)

Country	Policy types	Characteristics	Achievements or advantages	Demonstration or disadvantages
Copenhagen (Denmark)	Energy and buildings, transportation and urban planning, and Food waste and water	Focusing on combined heat and power, wind energy, building efficiency, and, most recently, electric vehicles, among numerous strategies to reduce carbon emissions and enhance the Danish economy (EESI, 1984)	• Combined heat and power (CHP), which provides 53 percent of Denmark's electricity and 80 percent of district heating • Wind turbines and photovoltaic systems. Wind Power accounts for 19% of total electricity production; • Green demonstration of buildings and improvement of energy systems such as strict building standards, a labeling system, and other residential and commercial efficiency policies; • Invest on electric vehicles and bikes. Charging stations and free parking for electric cars. The construction of bike lines and parking will be introduced • High taxes on energy use will be implemented • Recycle the waste (Only 3% of our waste goes to landfills)	It could be the first carbon neutral capital city, a model without obvious drawbacks

(continued)

Table 6.3 (continued)

Country	Policy types	Characteristics	Achievements or advantages	Demonstration or disadvantages
Helsinki (Finland)	Energy & Buildings, Transportation & Urban planning, Public participation, and behavior guidance	Most of emissions are from building heating, electricity consumption and transportation. The city would reduce emissions in transportation, urban construction, consumer procurement, household carbon footprint reduction (City of Helsinki, 2018)	• Increase the proportion of new energy vehicles. Development of heavy vehicle emission reduction technology to reduce emissions from the transport system • Control transportation, recycle waste and heat energy such as improve the energy consumption efficiency of building construction and use • Advocate green consumption. Guide market purchasing behavior and develop sharing and circular economy; • Replace coal power generation with new energy sources; • Enhance carbon sink through land use planning and green space maintenance • Citizen participation and establish a climate working group	The development of emissions and the progress are to be monitored only the GHG emissions in the area of the city. The Action Plan do not include the objectives about climate change or risks and the actions of centralised energy production (City of Helsinki, 2018)

(continued)

Table 6.3 (continued)

Country	Policy types	Characteristics	Achievements or advantages	Demonstration or disadvantages
Paris (France)	Energy and buildings, transportation and urban planning, food waste and water, and adaptation and implementation	Due to large share of nuclear and hydraulic production of electricity in particular, France has one of the lowest per capita CO_2 emission rates of any developed country (The French government, 2016)	• Large share of nuclear and hydraulic production of electricity • Rise of nuclear share's electricity generation • Crop-growing and livestock rearing • Energy transition for green growth act • Carbon capture and storage • Tax on fossil fuel	Emissions linked to household consumption remain high, given the greenhouse gas emissions generated by the production of imported goods
Singapore	Energy and buildings, transportation and urban planning, and adaptation and implementation	Weak climate target and needs to make additional policies to satisfy the increasing energy demand Emissions are dominated by the energy and industry sectors, representing 38% and 14%, respectively. The city's mitigation strategy is based on increasing energy and carbon efficiency, reducing carbon emissions in power generation, and developing low carbon technology	• Green buildings (LEDS builds, BCA Green Mark scheme in 2005, Singapore Green Building Masterplan (SGBMP) (Jain et al., 2020); • Transport Systems. Well-developed public rapid transit system. Private vehicle ownership restraint and usage management policies (Boey & Su, 2014) • Carbon tax on industrial facilities	• Increasing amount of GHG emissions. Singapore's GDP is increasing, and so are its absolute emissions • Lack of energy. Singapore lacks conventional energy resources and its alternative energy is disadvantaged (Su & Ang, 2020)

(continued)

Table 6.3 (continued)

Country	Policy types	Characteristics	Achievements or advantages	Demonstration or disadvantages
Tokyo (Japan)	Energy and buildings, food, waste and water, and adaptation and implementation	Long-term goal: 80% by 2050 The government's current policies are projected to meet its "Highly Insufficient" 2030 New climate commitment (NDC) target while promoting some positive policy development in the transport sector and fossil fuel power. The baseload plant will still account for a significant share of energy in the future (Timperley, 2018)	• The world's highest level of energy efficiency in steel and cement production • Public transport has a high share. Compared to other industrialised countries, its share stands at 46.7% of the total transportation network • Low carbon products. Producing and selling large numbers of products (hybrid automobiles and solar cells) • Low-energy vehicles • Commercial and household carbon emissions reduction • Support nuclear energy constantly • Expanded renewable power (i.e., solar, wind, and other non-hydro renewables) • Efficiently manages water resources • Climate laws, which help to guide the state, local governments and companies to develop emission reduction plans	• Easily affected by climate change and natural disasters (e.g., sea level rise, coastal erosion, more intense typhoon, and summer droughts) • Fossil fuel power plants would play an important role in Japan's energy mix in 2030 • Current energy strategy foresees a relatively large share of base load power plants of 46–48% in 2030 of total electricity production

6.5 Issues and Prospects: Transition to Sustainability

There has been much literature on the importance of climate change mitigation strategies and actions in regions and cities. A large number of variables have been proposed to support and justify these ideas. However, there are still some gaps that need to be further explored and further innovated in theory and practice. First, the causal relationship between local low carbon rhetoric and poor outcomes in cities and surrounding areas is ambiguous. The boundaries and key elements of effective and ineffective low carbon responses are unclear. Second, other than the interrelationships among key elements of low carbon, the relative importance of studies on top-down influences, choice of governance techniques, local politics, local authorities' institutional capacity, transnational municipal networks, and public engagement is not prominent (Lo, 2014). In addition, the temporal factors, especially the dynamic effects of long-term climate change, have been poorly researched. Longitudinal studies can help identify changes in urban low carbon responses and clarify the impact of time-related factors. Finally, there are few studies comparing countries at different levels of development. In this regard, we corroborate that low carbon cities' governance is likely to become more diversified in the future. We also note that many key innovations and leading experiments or experimental projects will come from cities in developing countries, notably China.

Low carbon cities attach equal importance to both adaptation and mitigation measures in the future. Firstly, there will be greater emphasis on strong economic policy tools and high-quality environmental planning. When solving the main contradiction of environmental governance, cities need to consider habitability and emission reduction. The green cities of the future will accept the vigorous development of carbon trading mechanisms and carbon trading markets (Liu et al., 2014). The government will further encourage all social stakeholders to participate in low carbon production and operation. The second is low carbon cities powered by technology. In recent years, new renewable energy technologies are being refined and optimised. Given the three major carbon emission sectors such as industry, transportation, buildings, and other high-polluting industries, we see advanced technologies will be actively introduced (Su et al., 2016). Finally, connectivity and synergy among agglomerations, metropolitan areas, cities, and suburban areas will promote low carbon and green development. For instance, small and medium-sized cities can relieve population pressure and economic, social, and environmental issues through coordination and trans-regional cooperation. Moreover, the technological connectivity and the active participation of the public are certainly examples that will continue to grow and become more effective in achieving low carbon targets and then zero-emission goals.

References

Adelaide City Council. (2015). *Carbon Neutral Adelaide Actin Plan 2016–2021.* Available from: https://d31atr86jnqrq2.cloudfront.net/docs/action-plan-carbon-neutralpdf.pdf?mtime=20190517094029&focal=none

Amsterdam City Council. (2019). *Amsterdam Climate Neutral 2050 Roadmap.* Available from: http://carbonneutralcities.org/wp-content/uploads/2019/12/Amsterdam-Climate-Neutral-2050-Roadmap_12072019-1.pdf

Azizalrahman, H., & Hasyimi, V. (2018). Towards a generic multi-criteria evaluation model for low carbon cities. *Sustainable Cities and Society, 39*, 275–282.

Boey, A., & Su, B. (2014). Low-carbon transport sectoral development and policy in Hong Kong and Singapore. *Energy Procedia, 61*, 313–317.

Building Research Establishment Environmental Assessment Methodology (BREEAM). (2021). *What is BREEAM?* Available from: https://www.breeam.com/

Butters, C., Cheshmehzangi, A., & Sassi, P. (2020). Cities, energy and climate: Seven reasons to question the dense high-rise city. *Journal of Green Building, 15*(3), 197–214.

Chavez, A., & Ramaswami, A. (2011). Progress toward low carbon cities: Approaches for transboundary GHG emissions' footprinting. *Carbon Management, 2*(4), 471–482.

Chen, J. G. (2011). Low-carbon city construction: International experience and China's policy choice. *Modern Property, 02*, 86–94.

Cheshmehzangi, A. (2016). China's New-Type urbanisation plan (NUP) and the foreseeing challenges for decarbonization of cities: A review. *Energy Procedia, 104*, 146–152.

Cheshmehzangi, A. (2020). Low carbon transition at the township level: Feasibility study of environmental pollutants and sustainable energy planning. *International Journal of Sustainable Energy*, 1–27.

Cheshmehzangi, A., & Butters, C. (Eds.). (2017). *Designing cooler cities: Energy, cooling and urban form: The Asian perspective.* Palgrave Macmillan.

Cheshmehzangi, A., & Dawodu, A. (2019). *Sustainable urban development in the age of climate change—People: Cure or curse.* Palgrave Macmillan.

Cheshmehzangi, A., Dawodu, A., & Sharifi, A. (2021). *Sustainable urbanism in China.* Routledge.

Cheshmehzangi, A., Xie, L., & Tan-Mullins, M. (2018). The role of international actors in low-carbon transitions of Shenzhen's international low carbon city in China. *Cities, 74*, 64–74.

Chester, M. V., Sperling, J., Stokes, E., Allenby, B., Kockelman, K., Kennedy, C., et al. (2014). Positioning infrastructure and technologies for low-carbon urbanization. *Earth's Future, 2*(10), 533–547.

City of Helsinki. (2018). *The carbon-neutral Helsinki 2035 action plan.* Available from: https://www.hel.fi/static/liitteet/kaupunkiymparisto/julkaisut/julkaisut/HNH-2035/Carbon_neutral_Helsinki_Action_Plan_1503019_EN.pdf

City of Paris. (2018). *The Paris climate action plan.* Available from: https://cdn.paris.fr/paris/2019/07/24/1a706797eac9982aec6b767c56449240.pdf

Climate Home News. (2020). *Denmark to phase out oil and gas production by 2050 in "watershed" decision.* Available from: https://www.climatechangenews.com/2020/12/04/denmark-phase-oil-gas-production-2050-watershed-decision/

Clean Energy Wire (CLEW). (2020). *Germany's climate action law.* Available from: https://www.cleanenergywire.org/factsheets/germanys-climate-action-law-begins-take-shape

Congress (USA). (2009). *American recovery and reinvestment act of 2009.* Available from: https://www.congress.gov/bill/111th-congress/house-bill/1/text

Dawodu, A., & Cheshmehzangi, A. (2017). Impact of floor area ratio (FAR) on energy consumption at Meso scale in China: Case study of Ningbo. *Energy Procedia, 105*, 3449–3455.

Department of Mineral Resources and Energy. (2019). Integrated Resource Plan (IRP2019). Available from: http://www.energy.gov.za/IRP/2019/IRP-2019.pdf

Dimitrov, R. S. (2010). Inside UN climate change negotiations: the Copenhagen conference. *Review of Policy Research,27*(6), 795–821.1.

Environmental and Energy Study Institute (EESI). (1984). *Denmark: A model of a low carbon, prosperous economy*. Environmental and Energy Study Institute. Available from: https://www.eesi.org/articles/view/denmark-a-model-of-a-low-carbon-prosperous-economy

European Commission (2020). *Hydrogen strategy in July 2020*. Available from: https://ec.europa.eu/info/news/adoption-eu-hydrogen-strategy-2020-jul-08_en

European Commission (2021). *Connecting Europe facility in telecom*. Available from: https://ec.europa.eu/digital-single-market/en/connecting-europe-facility-telecom

European Energy Research Alliance (EERA). (2020). *EC- new circular economy action plan*. Available from: https://www.eera-recyclers.com/news/ec-new-circular-economy-action-plan

Evans, J., & Karvonen, A. (2014). 'Give me a laboratory and I will lower your carbon footprint!'—Urban laboratories and the governance of low-carbon futures. *International Journal of Urban and Regional Research, 38*(2), 413–430.

Fang, J. Y., Wang, S. P., Yue, C., Zhou, J. L., Guo, Z. D., He, C. F., et al. (2009). Scenario analysis on the global carbon emissions reduction goal proposed in the declaration of the 2009 G8 summit. *Science in China, 52*(011), 1694–1702.

Federal Ministry for Economic Affairs and Energy. (2017). Renewable Energy Sources Act (EEG 2017). Available from: https://www.bmwi.de/Redaktion/EN/Downloads/renewable-energy-sources-act-2017.html

Fu, H. (2010). Overview of low-carbon economy research. *Asian Agricultural Research*, 2(1812-2016-143213), 65–74.

German Sustainable Building Council (DGNB). (2021). *The DGNB*. Available from: https://www.dgnb.de/en/index.php

Glaeser, E. L., & Kahn, M. E. (2010). The greenness of cities: Carbon dioxide emissions and urban development. *Journal of Urban Economics, 67*(3), 404–418.

Gu, C., Tan, Z., Liu, W., Yu, T., Han, Q., Liu, H., Dai, Y., Liu, Z., & Zheng, S. (2009). A study on climate change, carbon emissions and low-carbon city planning. *In Urban planning forum* (Vol. 3, pp. 38–45).

Harris, S., Weinzettel, J., & Levin, G. (2020). Implications of low carbon city sustainability strategies for 2050. *Sustainability*, 12.

Hasyimi, V., & Azizalrahman, H. (2018). A strategy-based model for low carbon cities. *Sustainability, 10*(12).

Ho, C. S., Matsuoka, Y., Simson, J., & Gomi, K. (2013). Low carbon urban development strategy in Malaysia—The case of Iskandar Malaysia development corridor. *Habitat International, 37*, 43–51.

Hong Kong Environment Bureau (HKEB). (2017). Hong Kong climate action blueprint 2030+. Retrieved February 11, 2021. Available from: https://www.legco.gov.hk/yr16-17/chinese/panels/ea/papers/ea20170626cb1-1371-2-c.pdf

Hongkong Council for Sustainable Development (HKCSD). (2019). *Long-term carbon reduction strategies—Public participation*. Retrieved February 11, 2021, Available from: https://www.gov.hk/sc/residents/government/publication/consultation/docs/2019/LDS.pdf

Jain, M., Siva, V., Hoppe, T., & Bressers, H. (2020). Assessing governance of low energy green building innovation in the building sector: Insights from Singapore and Delhi. *Energy Policy, 145*, 111752.

Kennedy, C. A., Ibrahim, N., & Hoornweg, D. (2014). Low-carbon infrastructure strategies for cities. *Nature Climate Change, 4*(5), 343–346.

Lee, C. T., Hashim, H., Ho, C. S., Fan, Y. V., & Klemes, J. J. (2017). Sustaining the low-carbon emission development in Asia and beyond: Austainable energy, water, transportation and low-carbon emission technology. *Journal of Cleaner Production, 146*(MAR.10), 1–13.

Liu, L. Q., Liu, C. X., & Gao, Y. G. (2014). Green and sustainable City will become the development objective of China's low carbon city in future. *Journal of Environmental Health Science and Engineering, 12*(1), 1–10.

Lo, K. (2014). Urban carbon governance and the transition toward low carbon urbanism: Review of a global phenomenon. *Carbon Management, 5*(3), 269–283.

Long, W., Bai, W., Liang, H., Fan, R., & Zhang, G. (2010). Urban form and energy vision in low-carbon cities. *Building Science, 26*(2), 13–23.

Ma, J., Zhou, L., & Li, W. (2010). Indicator system construction for urban low carbon economy development. *Science & Technology Progress and Policy, 027*(022), 165–167.

Milner, J., Davies, M., & Wilkinson, P. (2012). Urban energy, carbon management (low carbon cities) and co-benefits for human health. *Current Opinion in Environmental Sustainability, 4*(4), 398–404.

Organization for Economic Cooperation and Development (OECD). (2015). *Aligning policies for a low-carbon economy*. Available from: https://www.oecd.org/environment/Aligning-Policies-for-a-Low-carbon-Economy.pdf

Olazabal, M., & Pascual, U. (2015). Urban low carbon transitions: Cognitive barriers and opportunities. *Journal of Cleaner Production, 109* (DEC.16), 336–346.

Powering Past Coal Alliance (PPCA). (2021). *PPCA global summit 2021*. Available from: https://www.poweringpastcoal.org/summit-2021

Price, L., Zhou, N., Fridley, D., Ohshita, S., Lu, H., Zheng, N., & Fino-Chen, C. (2013). Development of a low-carbon indicator system for China. *Habitat International, 37*, 4–21.

Qiu, W. J., & Li, S. Y. (2011). Successful practice and beneficial enlightenment of low-carbon city construction at home and abroad. *Industry and Technology Forum, 08*, 7–9.

Raven, R., Sengers, F., Spaeth, P., Xie, L., Cheshmehzangi, A., & de Jong, M. (2019). Urban experimentation and institutional arrangements. *European Planning Studies, 27*(2), 258–281.

Rickerson, W., & Hughes, K. (2006). Municipal climate change initiatives and the policy framework for greenhouse gas reductions in New York City. In International Solar Cities Congress.

Romero-Lankao, P., Gurney, K. R., Seto, K. C., Chester, M., Duren, R. M., Hughes, S., et al. (2014). A critical knowledge pathway to low-carbon, sustainable futures: Integrated understanding of urbanization, urban areas, and carbon. *Earths Future, 2*(10), 515–532.

Scottish Government. (2010). *Low carbon Scotland: A behaviours framework*. Available from: https://www.gov.scot/binaries/content/documents/govscot/publications/strategy-plan/2013/03/low-carbon-scotland-behaviours-framework/documents/00415744-pdf/00415744-pdf/govscot%3Adocument/00415744.pdf

Scott, M. (2018). *Cities sign up to zero-carbon buildings by 2030*. Available from: https://www.forbes.com/sites/mikescott/2018/09/05/cities-sign-up-to-zero-carbon-buildings-by-2030/?sh=397672866ac4

Seoul Metropolitan Government. (2015). *Promise of Seoul: Taking actions against climate change*. Available from: https://www.ccacoalition.org/sites/default/files/resources/2015_Promise-of-Seoul_Seoul-Metropolitan-Government.pdf

Singapore Statutes. (2019). *Building control act*. Available from: https://sso.agc.gov.sg/Act/BCA1989?ValidDate=20190301&ProvIds=P1VI-

State of Green (2021). *Copenhagen—Carbon neutral by 2025*. Available from: https://stateofgreen.com/en/partners/city-of-copenhagen/solutions/copenhagen-carbon-neutral-by-2025/

Su, B., & Ang, B. W. (2020). Demand contributors and driving factors of Singapore's aggregate carbon intensities. *Energy Policy, 146*, 111817.

Su, M., Li, R., Lu, W., Chen, C., Chen, B., & Yang, Z. (2013). Evaluation of a low-carbon city: Method and application. *Entropy.*

Su, M., Zheng, Y., Yin, X., Zhang, M., Wei, X., Chang, X., & Qin, Y. (2016). Practice of low-carbon city in China: The status quo and prospect. *Energy Procedia, 88*, 44–51.

Tan, S., Yang, J., & Yan, J. (2015). Development of the low-carbon city indicator (LCCI) framework. *Energy Procedia, 75*, 2516–2522.

The City of Amsterdam. (2021). *Policy: Clean air*. Available from: https://www.amsterdam.nl/en/policy/sustainability/clean-air/

The French government. (2016). *Report of the implementation by France of sustainable development goals*. Available from: https://sustainabledevelopment.un.org/content/documents/10726Report%20SDGs%20France.pdf

The UK Department Trade. (2003). *Energy White Paper: Our energy future–creating a low carbon economy.* Available from: https://assets.publishing.service.gov.uk/government/uploads/system/uploads/attachment_data/file/272061/5761.pdf

The World Wide Fund (WWF) and CSUS (Chinese Society for Urban Studies). (2021). *Reducing carbon emissions in the UK*. From World Wide Fund for Nature. Retrieved February 11, 2021, Available from: http://webadmin.wwfchina.org/storage/files/2.%E9%9D%A2%E5%90%912035%E7%9A%84%E5%9F%8E%E5%B8%82%E4%BD%8E%E7%A2%B3-%E5%86%85%E6%96%870125.pdf

The World Wide Fund (WWF). (2020). *Beyond 2 °C—The long-term emission reduction actions of global cities in response to climate change.* Retrieved February 11, 2021. Available from: http://www.wwfchina.org/publication.php

The World Wide Fund (WWF). (2021). *Reducing carbon Emissions in the UK. From world wide fund for nature.* Retrieved February 11, 2021, Available from: https://www.wwf.org.uk/what-we-do/projects/reducing-carbon-emissions-uk

Timperley, J.(2018). The carbon brief profile: Japan. *CarbonBrief*. Available from: https://www.carbonbrief.org/carbon-brief-profile-japan

United Nations Environment Programme (UNEP). (2013). *Low carbon city: A guidebook for city planner and practitioners.* Available from: https://www.unep.org/resources/report/low-carbon-city-guidebook-city-planner-and-practitioners

United Nations Environment Programme (UNEP). (2020). *London aims to be a zero-carbon city by 2050.* Available from: https://www.unep.org/news-and-stories/story/london-aims-be-zero-carbon-city-2050

van Doren, D., Driessen, P. P., Runhaar, H., & Giezen, M. (2018). Scaling-up low-carbon urban initiatives: Towards a better understanding. *Urban Studies, 55*(1), 175–194.

Wu, J., & Zhu, Q. T. (2020). An analysis of the role of cities in global climate governance in the post-Paris Era. *Environmental Protection,48*; 676 (05), 14–19.

Wu, X. (2014). International experience and path selection of low-carbon city construction. *Regional Economic Review, 01*, 40–47.

Xia, K. B. (2008). Develop low carbon economy to achieve sustainable urban development. *Environmental Protection, 03*, 33–35.

Xin, Z. P., & Zhang, Y. T. (2008). Low Carbon Economy and Low Carbon City. *Urban Studies, 15*(004), 98–102.

Yokota, M. (2010). Isocarp low carbon cities -45th world congress in porto, PORTUGAL. *City Planning*, 59.

Zeng, S. H., & Zhang, S. A. (2011). Literature review of carbon finance and low carbon economy for constructing low carbon society in china. *Low Carbon Economy, 2*(1), 15–19.

Chapter 7
Overview of China's Low Carbon Progress: Policies and Initiatives

7.1 China's Early Exploration of Low Carbon Development Before 2005

In this chapter, we provide an overview of China's low carbon progress in multiple stages. First, we look into China's early exploration before 2005, and then the main years of the 11th and 12th Five-Year Plans (FYPs). We then briefly evaluate the more recent progress of the 13th FYP before exploring the next steps of the 14th FYP, 2030 targets, and the 2060 plan. By exploring these stages, we look into policies and initiatives and progress and actions that have taken place to date. In the later chapters, we provide more details and move towards future directions.

Before low carbon was proposed, China had already made exploration on energy conservation and environmental protection. China began to implement an open-door policy in 1979 and experienced spectacular economic growth (Zhang, 2010). The rapid age of China's urbanisation led to the development of new directions and strategies that fast-forward industrial production, construction, and physical infrastructure development (Cheshmehzangi, 2016; Cheshmehzangi et al., 2021). Although the "Environmental Protection Law of the PRC" (NPC, 1979) was formally promulgated, China's environment saw unprecedented pollution in the 1980s. This was the starting point of China's rapid urbanisation. In the absence of a high level of scientific and technological development, it was mainly up to the government to adopt administrative, legal, and economical means to strengthen its environmental management. The Provisional Regulations on the Management of Energy Conservation promulgated by the State Council in 1986 is China's first administrative regulation on energy conservation. It was replaced by "the Law on Energy Conservation of the PRC" (NPC, 1997) as the first energy conservation law. These are amongst China's early exploration of low carbon development.

China's sustainable development began in the 1990s and implemented policies to reduce GHG emissions in the context of global environmental issues (Liu & Qin, 2016). Through "China's Agenda 21: White Paper on China's Population, Environment, and Development in the 21st Century", China has formulated an action

A. Cheshmehzangi and H. Chen, *China's Sustainability Transitions*,
https://doi.org/10.1007/978-981-16-2621-0_7

plan for implementing the strategy of sustainable development (The State Council, 1994). Then, China made the sustainable development strategy a national strategy in the ninth five-year plan in 1995. In 2001, China joined the World Trade Organization (WTO), which led to further development strategies. At that time, the environmental and resource problems brought about by the rapid economic development were becoming increasingly prominent. In order to transform and upgrade its industries, China had implemented a series of policies to promote green development such as the "Cleaner Production Promotion Law of the PRC" (The State Council, 2002), the "Measures for the Administration of Energy Efficiency Labels" (The State Council, 2004), and the "Interim Provisions on Promoting Industrial Structure Adjustment" (The State Council, 2005a). In the context of increasingly fierce market competition, energy consumption was growing rapidly. Therefore, in 2004 and 2005, the State Council issued the "Notice on the implementation of resource conservation activities" and the "Notice on the recent key work of building a conservation-oriented society", respectively. China's extensive mode of economic growth had led to problems such as high resource consumption, waste, and serious environmental pollution, which then led to the issuance of "Several Opinions of the State Council on Speeding up the Development of Circular Economy" (The State Council, 2005c). At this stage, the Chinese government advocated raising resource utilisation efficiency and saving resources in production, construction, mobility, and consumption.

7.2 Low Carbon Development During 11th and 12th Five-Year Plans Period

The Chinese government included climate change as a major issue in the economic and social development plan. In 2006, China set a mandatory target of reducing energy consumption per unit of GDP by 20% by 2010 from the 2005 level. In 2007, China became the first developing country to formulate and implement a national plan to address climate change. In 2009, China set an action target of reducing greenhouse gas emissions per unit of gross domestic product by 40–45% from 2005 levels by 2020. To achieve these goals, China formulated a series of major policies and measures for climate change mitigation and adaptation during the 11th FYP (2006–2010). In the 12th FYP (2011–2015), China established the policy orientation of promoting green and low carbon development and clearly set out the goals and tasks for addressing climate change.

During this period, China also carried out (a series of) pilot projects on low carbon provinces and cities and made some progress. In 2010, China began to carry out pilot work in Guangdong, Liaoning, Hubei, Shaanxi, and Yunnan provinces. This pilot work also included eight cities of Tianjin, Chongqing, Shenzhen, Xiamen, Hangzhou, Nanchang, Guiyang, and Baoding (NDRC, 2010). It included the formulation of low carbon development plans and supporting policies to support low carbon and green development. While accelerating the establishment of an industrial system featuring low carbon emissions, a greenhouse gas emission data statistics and management system were established. In addition, low carbon green lifestyle and consumption patterns were advocated. The "National New-type Urbanisation Plan" (2014–2020) proposed ecological civilization (Cheshmehzangi, 2014; Deng & Cheshmehzangi, 2018), green and low carbon urbanization (The State Council, 2014). Those efforts vigorously promoted the transformation of the economic development mode (Xie et al., 2020), advanced the construction of key low carbon development projects, and developed low carbon industries (Cheshmehzangi, 2020) to promote green and low carbon development.

7.2.1 Policies and Actions

The 11th FYP was the period when China began to build a moderately prosperous society, which can develop energy conservation and emission-reduction (Hu et al., 2011). During this period, China introduced a series of policies to promote low carbon development and put in place a preliminary energy conservation management system. Consequently, China made vast investments in three primary energy-related areas of new energy, renewable energy and energy efficiency. In 2006, the "Eleventh Five-Year Plan" (NPC, 2006), for the first time, put forward the concept of resource conservation as a basic state policy and set forth targets for energy conservation and emission reduction. The release of the "Renewable Energy Law" (NEA, 2006) and the "Medium and Long-Term Development Plan for Renewable Energy" (NDRC, 2007) marked the beginning of China's focus on energy structure (Cheshmehzangi, 2020). According to China's National Climate Change Program (The State Council, 2007), the country's first global warming policy initiative, the government adopted measures ranging from laws, economy, administration, and technology, combined to reduce GHG emissions. Soon after, pricing, financial, fiscal, and tax policy tools were actively used in conjunction with government investment. Government-led administrative measures and incentives then played a leading role in energy conservation and carbon reduction. In January 2008, China's National Development and Reform Commission (NDRC) and the World Wildlife Fund (WWF) jointly selected Shanghai and Baoding as pilot low carbon cities. As a result, "White paper: China's policies and actions on climate change" (NPC, 2008) was released in October 2008, which proposed to adjust the economic structure to promote the optimising and upgrading of the industrial structure. This was mainly developed to make great efforts to save energy, raise energy efficiency, and develop a recycling economy. Somehow, it could be argued that this was the starting point for discussions on developing towards a circular economy model.

In the 12th FYP (2011–2015), China continued to pursue the concept of low carbon economic development. In this important period, low carbon became one of the core concerns of urbanisation, which already faced a more stabilised pace and pattern than before. China took energy intensity per unit of GDP as a binding target and called for appropriate total energy consumption control. To ensure the achievement of energy conservation and emission reduction targets, China issued the "Work Plan for Greenhouse Gas Emission Control during the 12th Five-Year Plan Period" (The State Council, 2011a), setting out the overall requirements and key tasks for controlling greenhouse gas emissions. The "12th Five-Year Plan for Energy Conservation and Emission Reduction" (The State Council, 2012) laid out an overall plan for energy conservation during the 12th FYP period. Since 2014, China has actively taken measures to tackle climate change, such as the "National Climate Change Plan" (2014–2020) and "China's Policies and Actions on Climate Change" (NDRC, 2015). At the same time, China actively promoted international exchanges and cooperation on climate change, issued joint statements on climate change with many other countries. This led to experimental projects of low carbon development with multiple partners (Cheshmehzangi et al., Cheshmehzangi, Xie, et al., 2018) or through bilateral and multilateral cooperation.

7.2.2 Progresses and Changes

During the 11th FYP period, China transformed economic development mode to mitigate or adapt to climate change (The State Council, 2011b). The first measure was the optimisation of industrial structure. The following changes were energy conservation, developing low carbon energy, controlling non-energy-related GHG emissions, increasing carbon sink, and promoting low carbon development in localities. On the contrary, the adaptation mainly focused on the agricultural sector, water resources, marine resources, public health, and meteorology (as shown in Table 7.1). With the efforts of all sectors, China has achieved the energy conservation targets set out in the 11th FYP. Energy consumption per unit of GDP was reduced by 19.1% over 2005, and carbon dioxide emissions were reduced by 1.46 billion tons. Over the same period, China's national economy grew at an average annual rate of 11.2%, while its energy consumption grew by only 6.6% (China Daily, 2011). China shut down an outdated iron-making capacity of 122 million tons, a steel production capacity of 69.69 million tons, cement production capacity of 330 million tons, and small thermal power units generating 72 million kilowatts. In terms of clean and new energy, the total installed capacity of hydropower in 2010 increased from 110 million kilowatts in 2005 to 200 million kilowatts; the installed capacity of nuclear power reached 10.82 million kilowatts; the installed capacity of wind power increased from 1.3 million to 40 million kilowatts; and the installed capacity of photovoltaic power generation increased from less than 100,000–600,000 kw. The forest coverage rate increased from 18.2% in 2005 to 20.36% in 2010 (CMA, 2011).

During the 12th FYP period, the Chinese government achieved positive results in mitigating climate change. The primary achievements were summarised as adjusting the industrial structure ("Catalogue for Guiding Industry Restructuring" 2011 Version and 2013 Amendment), optimising the energy structure ("the strategic action plan for energy development 2014–2020"), saving energy, and improving

Table 7.1 Mitigating and adapting to climate change progresses in China (Drawn by the Authors, using the translated data from the State Council, 2011b)

Mitigating to climate change	
Optimising industrial structure	Reforming and upgrading traditional industries
	Fostering and strengthening strategic and newly emerging industries
	Accelerating the development of the service industry
Energy conservation	Enhancing target responsibility assessment
	Promoting energy conservation in key fields
	Promoting energy-saving technology and products
	Developing a circular economy
	Promoting energy conservation market mechanism
	Improving related standards
	Incentive policies
Developing low carbon energy	Accelerating the development of natural gas and other clean resources
	Proactively developing and utilising non-fossil energy
Controlling non-energy-related greenhouse gas emission	Enhanced control over GHG emission in industrial and agricultural production, waste disposal and other fields
Increasing carbon sink	Ecological protection projects
	Improving farmland and grassland carbon sinks
Promoting low carbon development in localities	Low carbon pilot projects in selected provinces and cities
Adapting to climate change	
Agriculture	Consolidated farmland and water conservancy infrastructure
Water resources	National comprehensive plan for water resources
	Seven major river basins' flood control plan
	National mountain torrent disaster prevention and control plan
	National plan to guarantee the safe supply of drinking water to urban dwellers
	National plan for the eco-protection of major rivers and lakes

(continued)

Table 7.1 (continued)

Marine resources	Strengthened the construction of a marine meteorological observation network
	Started redefining the national and provincial marine functional zoning
Public health	The national health emergency response plan against natural disasters
	The emergency response plan against high-temperature heat-stroke
	The national environment and health action plan (2007–2015)
Meteorology	The weather research plan (2009–2014), climate research plan (2009–2014), applied meteorology research plan (2009–2014) and comprehensive Meteorological observation research plan (2009–2014)
	Climate research plan (2009–2014)
	Meteorology research plan (2009–2014)
	Comprehensive meteorological observation research plan (2009–2014)

energy efficiency ("the 2014–2015 Action Plan for Energy Conservation, Emission Reduction and Low carbon Development"), and controlling GHG emissions from non-energy activities and increasing carbon sink ("the Outline of the National Afforestation Plan 2011–2020") (NDRC, 2016). During this five-year period, China cut carbon dioxide emissions per unit of GDP for energy activities by 20%, exceeding the binding target of a 17% reduction (Fig. 7.1).

In 2009, just before the UN's Copenhagen Climate Change Conference, China announced that by 2020 it would lower carbon dioxide emissions per unit of GDP by 40–45% from the 2005 level, increase the share of non-fossil fuels in primary energy consumption to about 15%, and increase the forested area by 40 million hectares and the forest stock volume by 1.3 billion cubic meters compared to the 2005 levels (Hu et al., 2011). China submitted its Intended Nationally Determined Contributions (INDC) to the UN Framework Convention on Climate Change (UNFCCC) in 2015 and promised to cut its carbon emissions per unit of GDP by 60–65% from 2005 levels by 2030, increase non-fossil fuel sources in primary energy consumption to about 20%, and peak its carbon emissions (ECNS, 2015).

During the 12th FYP period, China has initiated carbon emission trading pilots in seven provinces and cities and low carbon development pilots in 42 provinces and cities. This pilot work was conducted to explore a new mode of low carbon development consistent with its prevailing national circumstances. The Switzerland-China Low Carbon Cities Project was launched in June 2010. In this project, eight cities were selected as pilot cities based on multiple criteria of city management, low carbon economy , transportation, and green building (Zhou, 2012). In 2012, the

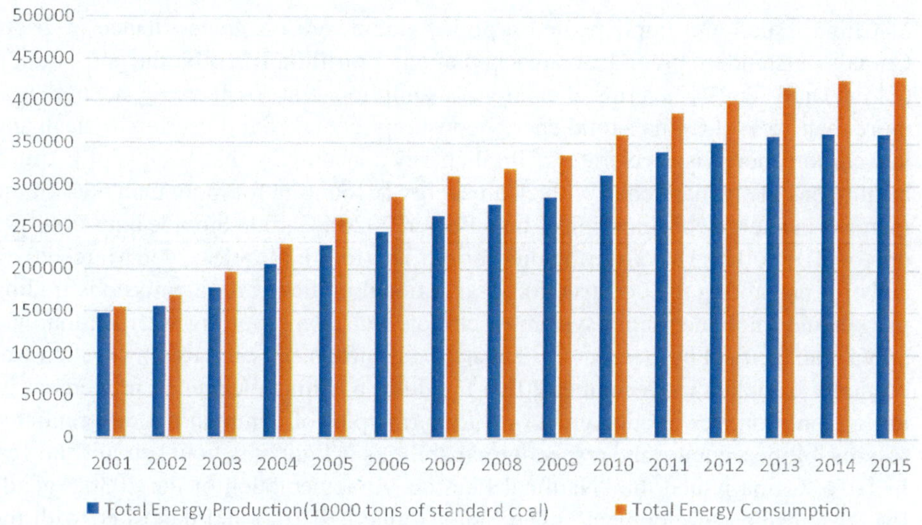

Fig. 7.1 China's total energy production and consumption (Drawn by the Authors, data from Gao, 2020, and China Energy Statistical Yearbook)

NDRC issued "Interim measures for voluntary greenhouse gas emissions trading management" (NDRC, 2012), introducing a voluntary emissions trading system in China. Subsequently, Beijing, Shanghai, Tianjin, Guangdong, and other provinces and cities set up carbon emission trading centers one after another to carry out pilot projects of mandatory trading of carbon emission rights. In June 2013, Shenzhen took the lead in launching carbon emission trading, followed by Shanghai and Beijing at the end of November 2013 (Zhao et al., 2017), which means the carbon trading market has entered the substantive operation stage.

7.3 Recent Progress: Low Carbon Development in the 13th Five-Year-Plan

During the 13th FYP (2016–2020), China's GHG emissions have been effectively controlled (Cheshmehzangi et al., Cheshmehzangi, Li, et al., 2018). By the end of 2019, carbon intensity had fallen by 18.2% from 2015, meeting the binding targets set for the 13th FYP ahead of schedule. Carbon intensity was 48.1% lower than in 2005, and non-fossil energy accounted for 15.3% of energy consumption. Installed renewable energy capacity will grow by about 12% annually. China carried out pilot projects to adapt to climate change in 28 cities and carried out pilot projects to build low carbon cities in 81 cities in six provinces and autonomous regions in three batches (China Economic Net, 2020). Over the last five years, the share of clean energy consumption has risen from 19.1 to 24.3%. After the People's Bank of China issued and implemented a policy framework on green finance in 2016, China's outstanding green loans reached about 12 trillion RMB by the end of 2020 (China Daily, 2021). In terms of energy consumption, non-fossil energy accounts for more than 15% of China's total energy consumption, and the transition to clean and low carbon energy is accelerating. Total energy consumption has been kept within 5 billion tons of standard coal, with clean energy accounting for more than 65% of the increase in energy consumption (China Power, 2020). Carbon dioxide emissions per unit of GDP fell by 18.5% during this period. The 13th FYP, released in March 2016, calls for promoting the construction of a national unified carbon emissions trading market and implementing a system of carbon emission reporting, verification, and quota management by key sectors. The national carbon emission trading system was officially launched in December 2017. In addition, China pledges to implement its mitigation promises in adherence with the principles of common but differentiated responsibilities, equity, and respective capabilities in light of national circumstances. In 2016, China issued the "National Plan on Implementation of the 2030 Agenda for Sustainable Development" (The State Council, 2016a) and deposited with the United Nations the ratification instrument of the Paris Agreement. China announced the goal of increasing nationally determined contributions, peaking carbon dioxide emissions by 2030 and becoming carbon neutral by 2060. ECNS summarised the highlights of China's green development during the 13th FYP period, including the following (ECNS, 2020a):

- "Carbon dioxide (CO_2) emissions per unit of the gross domestic product fell by 18.2% by the end of last year from 2015, completing ahead of schedule the target set over the five years.
- In 2019, the energy consumption per unit of value-added industrial output by major firms dropped by more than 15% compared with 2015, equivalent to saving 480 million tonnes of standard coal.
- China has reduced outdated steel capacity total 200 million tonnes and upgraded coal-fired power plants and steel factories to ensure ultra-low emissions."

In 2019, the average consumption rate of wind energy in China was 96%. The average consumption rate of solar photovoltaic hitting 98%, and the average consumption rate of water energy in major river basins reaching 96% (ECNS, 2020b). The electrification rate of China's railways reached 71.9%. More than 400,000 buses and 430,000 trucks use new energy, 180,000 natural gas vehicles, and 290 liquefied natural gas (LNG) ships (ECNS, 2020c).

During this period, many central government policies mentioned low carbon development. "Notice of the State Council on Issuing the Work Plan for Greenhouse Gas Emission Control during the 13th Five-Year Plan Period" (The State Council, 2016b) proposed that by 2020, carbon dioxide emissions per unit of the gross domestic product should be cut by 18% from 2015 levels. "Key Energy-Efficient Technology Promotion Catalogues" (MEE, 2017) covers a total of 27 national key low carbon technologies in non-fossil energy, fuel, and raw material substitution, process, and other fields such as non-carbon dioxide emission reduction, carbon capture, utilisation and storage, and carbon sink. "Three-year plan of action for winning the war to protect blue skies" (The State Council, 2018) stressed building a clean, low carbon, and efficient energy system and a green transportation system.

7.4 Steps Ahead to Peak: During the 14th Five-Year Plan

China pledges to lower carbon dioxide emissions per unit of GDP by over 65% from the 2005 level, raise the share of non-fossil energy in primary energy use to around 25% by 2030. During the 14th FYP (2021–25) period, China will control the total and intensity of energy consumption, reduce the use of fossil fuels, and promote low carbon transformation in the industry, construction, transportation, and other sectors. On February 22, 2021, China issued the "Guiding Opinions on Accelerating the Establishment and Improvement of a Green and Low carbon Circular Development Economic System" (The State Council, 2021a), which set an ambitious and comprehensive direction to put China on a greener development path. China plans to cut energy consumption per unit of GDP by 13.5% and carbon dioxide emissions by 18%, according to the Government Work Report (The State Council, 2021b). The new guidelines were delivered by Premier Li Keqiang on March 5, 2021. The outline of the 14th FYP for national economic and social development and the long-range objectives through the year 2035 (the fourteenth Five-Year Plan and vision for 2035) proposed that during this FYP period, the urbanisation rate will be raised to 65%, the forest coverage rate increase to 24.1%, and the share of non-fossil energy in total energy consumption be raised to around 20% (NPC, 2021).

In the next five years, China will further transform its energy consumption structure by promoting the wider use of clean and renewable energy during this period. Measures to build a modern energy system include developing non-fossil energy sources, increasing the scale of wind power and photovoltaic power generation, controlling the development of coal power, promoting the substitution of electricity for coal, orderly lifting market access for oil and gas exploration and development, and intelligent upgrading of power grid infrastructure. The 14th FYP also puts forward some construction of low carbon city initiatives like developing green urban rings, corridors, wedges, and greenways, promoting ecological restoration and functional improvement, giving priority to the development of urban public transport, a slow-traffic network of bicycle lanes, pedestrian paths, smart construction, and promoting the use of green building materials, prefabricated buildings, and steel-structure housing, etc. China needs more finance stimulus packages to facilitate the low carbon transition of various sectors and support low carbon and green transformation projects. China will also improve the system of green financing standards.

Some regions in China will take the lead in peaking carbon emissions ahead of schedule (China Daily, 2021). More than 80 pilot low carbon cities have set the goal of reaching the peak in their studies, and 42 of them aim to reach the peak by 2025 (ESCN, 2021). At the provincial level, Shanghai, Fujian, Hainan, and

Qinghai proposed to meet the national peak figures before the proposed 2030 timeline. Furthermore, several provinces, including Tianjin, Shanghai, Hebei, Shanxi, Jiangsu, Anhui, Fujian, Jiangxi, Shandong, Henan, Shaanxi, Liaoning, Hubei, Hainan, Sichuan, Gansu, and Tibet, proposed to make a carbon dioxide emission plan by 2021. While the 14th FYP has just started, we see tremendous progress and plans set to achieve the 2030 carbon targets. As the book is completing, we anticipate further development on these initiatives, and we expect to see new guidelines and reports in the coming months and years.

7.5 The Definition of Carbon Neutral: To Meet the 2060 Target

There are some common international standards for carbon neutral certification. ISO 14064, which is a standard mainly for the quantification of GHG, was proposed by the International Organization for Standardization (ISO) in 2006 (ISO, 2006). PAS 2060 is the first international standard in the world to propose carbon neutral certification. It was published by the British Standards Institute (BSI) in 2010 and can be applied to all types of organisations (BSI, 2021). It proposed the quantification, reduction, and compensation of GHG emissions to achieve and implement the regulations. ISO has set up a 'Working Group on Carbon Neutralization' to launch the development of ISO 14068, an international standard related to carbon neutrality (ISO, 2021). The standard is expected to be finalised and released in 2023. These standards could help provide a unified approach and principles for achieving carbon neutrality and support the better use of carbon–neutral targets and specifications by countries in developing climate change plans, strategies, and programmes. The carbon neutral implementation procedures mentioned in the "Implementation Guideline on Carbon Neutral for Large-scale Activities (Trial)" (MEE, 2019) include five parts: carbon neutral planning, implementation of emission reduction actions, quantification of greenhouse gas emissions, carbon neutral activities, and carbon neutral evaluation. We use models under different scenarios to predict China's path to becoming carbon neutral. The carbon emissions by different sectors are shown in Fig. 7.2. The international path of 2 °C would not be carbon–neutral before 2060. The 1.5 °C path is expected to be carbon–neutral by 2055. China's policy scenario is somewhere between them so that this ambitious goal can be achieved by 2060. Some research institutions have also confirmed similar conclusions.

In sum, this chapter provided an overview of China's low carbon progress, and specifically from the perspectives of policies and initiatives. In recent FYPs, We see emerging areas of progress and agenda that have enabled experimental projects, pilot cases, regional/provincial agendas, and local/municipal targets. The 14th and 15th FYPs will be crucial to the next steps of meeting the 2030 targets, which are then the foundation for the 2060 carbon neutrality plan. As we highlighted, the prolonged process has gone through several stages of transitions, highlighting the importance

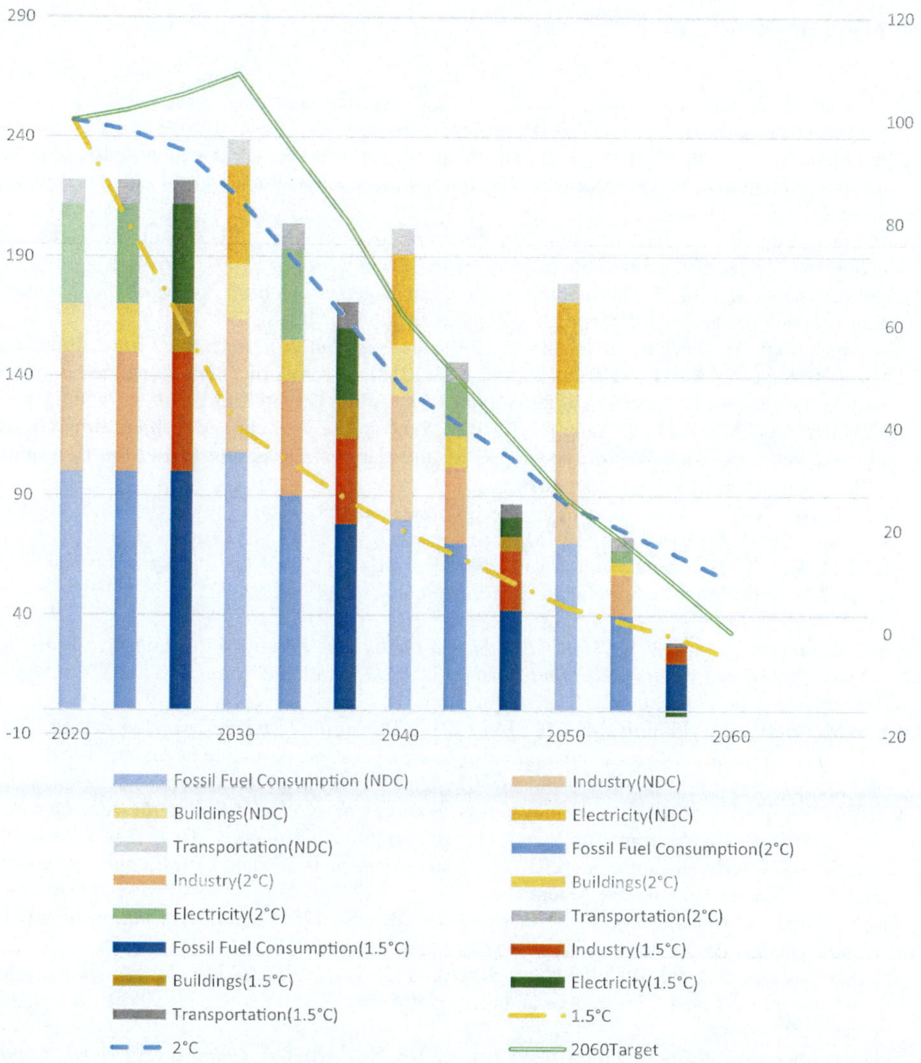

Fig. 7.2 China's carbon emissions by sector and total emissions under different scenarios (Unit: 100 million tons of CO_2) (Drawn by the Authors, data from ICCSD, 2020, GEIDCO, 2021

of systematic progression based on a procedural approach. In the next chapter, we delve into more low carbon practices and lessons that could lead to ultimate carbon neutrality goals. In doing so, the next chapter will be a complimentary one to this chapter, with the aim to cover key aspects of low carbon transitions in China as well as practices that apply to the next steps.

Added Note As the 14 FYP is just started, we anticipate more updates in the coming months and years. Therefore, we can only include guidelines and reports up to the end of March 2021.)

References

British Standards Institution (BSI). (2021). *PAS 2060 carbon neutrality*. Available from: https://www.bsigroup.com/en-GB/PAS-2060-Carbon-Neutrality/

Cheshmehzangi, A. (2014). The urban and urbanism: China's new urbanization and approaches towards comprehensive development. *The International Journal of Interdisciplinary Environmental Studies, 8*(3–4), 1–12.

Cheshmehzangi, A. (2016). China's new-type urbanisation plan (NUP) and the foreseeing challenges for decarbonization of cities: A review. *Energy Procedia, 104*, 146–152.

Cheshmehzangi, A. (2020). Towards urban energy management: A brief review of China's recent transition in urban energy demand. *Low Carbon Economy, 11*(1), 1–24.

Cheshmehzangi, A., Dawodu, A., & Sharifi, A. (2021). *Sustainable urbanism in China*. Routledge.

Cheshmehzangi, A., Xie, L., & Tan-Mullins, M. (2018). The role of international actors in low carbon transitions of Shenzhen's International Low Carbon City in China. *Cities, 74*, 64–74.

Cheshmehzangi, A., Li, H., & Yang, X. (2018b). Low carbon and climate resilient urban development in China. ADB Report, developed for the National Development Reform Committee (NDRC), Beijing, China, October 2018.

China Daily. (2011). *China's climate change efforts*. Available from: http://www.chinadaily.com.cn/cndy/2011-11/23/content_14145201.htm

China Daily. (2021). *China's new five-year blueprint paves way for 2060 carbon-neutrality*. Available from: https://www.chinadaily.com.cn/a/202103/09/WS60475593a31024ad0baae061.html

China Economic Net. (2020). *China has achieved the binding target for the 13th Five-Year Plan ahead of schedule by cutting carbon intensity by 18.2%*. Available from: http://www.ce.cn/cysc/stwm/gd/202010/22/t20201022_35924000.shtml

China Meteorological Administration (CMA). (2011). *During the 11th Five-Year Plan period, China has achieved remarkable results in addressing climate change*. Available from: http://www.cma.gov.cn/2011xwzx/2011xqhbh/2011xdtxx/201111/t20111109_151385.html

China Power. (2020). *Review the achievements in energy development during the 13th Five-Year Plan period*. Available from: http://www.chinapower.com.cn/flfd/hyyw/20201224/39784.html

Deng, W., & Cheshmehzangi, A. (2018). *Eco-development in China: Cities, communities and buildings*. Singapore: Palgrave Macmillan.

English-language website of China News Service (ECNS). (2015). *China in action to address climate change*. Available from: http://www.ecns.cn/voices/2015/11-28/190499.shtml

English-language website of China News Service (ECNS). (2020a). *Highlights of China's green development over past 5 years*. Available from: http://www.ecns.cn/news/2020-10-27/detail-ihacfvrp0929190.shtml

English-language website of China News Service (ECNS). (2020b). *China gives priority to non-fossil energy development: White paper*. Available from: http://www.ecns.cn/news/politics/2020-12-21/detail-ihaexsqc6150297.shtml

English-language website of China News Service (ECNS). (2020c). *China's transport sector seeks low-carbon development: White paper*. Available from: http://www.ecns.cn/news/2020-12-22/detail-ihaexsqc6150512.shtml

ESCN. (2021). *A brief analysis of the measures taken by some countries and Chinese local governments, industries and enterprises to implement carbon peak and carbon neutral targets*. Available from: http://www.escn.com.cn/news/show-1178488.html

Gao, C. (2020). Research on prediction of china's total energy production based on MCMC method. *Journal of Physics Conference Series, 1549*, 042057.

Global Energy Interconnection Development and Cooperation Organization (GEIDCO). (2021). *Report on China carbon neutrality before 2060*. Available from: https://m.book118.com/html/2021/0319/6243241121003122.shtm

Hu, Z., Yuan, J., & Hu, Z. (2011). Study on China's low carbon development in an economy–energy–electricity–environment framework. *Energy Policy, 39*(5), 2596–2605.

Institute of Climate Change and Sustainable Development, Tsinghua University (, ICCSD, &). (2020). China low-carbon development strategy and transformation pathways. *China Population, Resources and Environment, 30*(11), 1–25.

International Standardization Organization (ISO). (2006). *ISO 14064-1:2006 greenhouse gases*. Available from: https://www.iso.org/standard/66453.html

International Standardization Organization (ISO). (2021). *ISO/WD 14068 greenhouse gas management and related activities—Carbon neutrality*. Available from: https://www.iso.org/standard/43279.html

Liu, W., & Qin, B. (2016). Low carbon city initiatives in China: A review from the policy paradigm perspective. *Cities, 51*, 131–138.

Ministry of Ecology and Environment of the PRC (MEE). (2017). *Key energy-efficient technology promotion catalogues*. 国家重点节能低碳技术推广目录>. Available from: http://www.mee.gov.cn/ywgz/ydqhbh/wsqtkz/201904/t20190419_700394.shtml

Ministry of Ecology and Environment of the PRC (MEE). (2019*). Implementation guideline on carbon neutral for large-scale activities (Trial)*. 大型活动碳中和实施指南 (试行). Available from: http://www.mee.gov.cn/xxgk2018/xxgk/xxgk01/201906/t20190617_706706.html

National Development and Reform Commission (NDRC). (2004). *Measures for the administration of energy efficiency labels*. 能源效率标识管理办法>. Available from: https://www.ndrc.gov.cn/xxgk/zcfb/fzggwl/201603/W020190905495017011772.pdf

National Development and Reform Commission (NDRC). (2007). *Medium and long-term development plan for renewable energy*. 可再生能源中长期发展规划>. Available from: http://www.nea.gov.cn/131053171_15211696076951n.pdf

National Development and Reform Commission (NDRC). (2010). *Notice on pilot work of low-carbon provinces, regions and cities*. 关于开展低碳省区和低碳城市试点工作的通知>. Available from: http://www.gov.cn/zwgk/2010-08/10/content_1675733.htm

National Development and Reform Commission (NDRC). (2012). *Interim measures for voluntary greenhouse gas emissions trading management*. 温室气体自愿减排交易管理暂行办法>. Available from: http://www.hbets.cn/wcs/Upload/201711/5a0ba9dd583c0.pdf

National Development and Reform Commission (NDRC). (2015). *China's policies and actions on climate change*. Available from: http://www.cma.gov.cn/en2014/news/News/201511/P020151120624172083192.pdf

National Development and Reform Commission (NDRC). (2016). *China's policy and actions on climate change 2016*. 中国应对气候变化的政策与行动2016年度报告>. Annual Report. Available from: http://www.gov.cn/xinwen/2016-11/07/content_5129521.htm

The National Energy Administration (NEA). (2006). *The renewable energy law of the PRC*. 中华人民共和国可再生能源法>. Available from: http://www.nea.gov.cn/2017-11/02/c_136722869.htm

The National People's Congress (NPC). (1979). *Environmental protection law of the PRC*. 中华人民共和国环境保护法>. Available from: http://www.npc.gov.cn/wxzl/gongbao/2000-12/10/content_5004381.htm

The National People's Congress (NPC). (1997). *The law on energy conservation of the PRC*. 中华人民共和国节约能源法>. Available from: http://www.npc.gov.cn/npc/c12435/201811/045c859c5a31443e855f6105fe22852b.shtml

The National People's Congress (NPC). (2006). *The Eleventh Five-Year Plan of the PRC*. 中华人民共和国国民经济和社会发展第十一个五年规划纲要>. Available from: http://www.npc.gov.cn/wxzl/gongbao/2006-03/18/content_5347869.htm

The National People's Congress (NPC). (2008). *White paper: China's policies and actions on climate change*. 白皮书:中国应对气候变化的政策与行动>. Available from: http://www.npc.gov.cn/zgrdw/npc/zxft/zxft8/2009-08/24/content_1515034.htm

The National People's Congress (NPC). (2021). *The fourteenth Five-Year Plan and vision for 2035 of the PRC*. 中华人民共和国国民经济和社会发展第十四个五年规划和2035年远景目标纲要>. Available from: http://www.npc.gov.cn/npc/kgfb/202103/bf13037b5d2d4a398652ed253cea8eb1.shtml

The State Council of the PRC. (1994). *China's agenda 21: White Paper on China's population, environment and development in the 21st Century*. 中国21世纪议程—中国21世纪人口,环境与发展白皮书>. Available from: http://risd.ccnu.edu.cn/__local/F/0B/12/5A470E3D188A9E51A3CC43E199E_DDCD31B0_60327.pdf

The State Council of the PRC. (2002). *Cleaner production promotion law of the PRC*. 中华人民共和国清洁生产促进法>. Available from: http://www.gov.cn/gongbao/content/2002/content_61640.htm

The State Council of the PRC. (2004). *Notice on the implementation of resource conservation activities*. 关于开展资源节约活动的通知>. Available from: http://www.gov.cn/xxgk/pub/govpublic/mrlm/200803/t20080328_32367.html

The State Council of the PRC. (2005a). *Interim provisions on promoting industrial structure adjustment*. 促进产业结构调整暂行规定>. Available from: http://www.gov.cn/zhengce/content/2008-03/28/content_2020.htm

The State Council of the PRC. (2005b). *Notice on the recent key work of building a conservation-oriented society*. 关于做好建设节约型社会近期重点工作的通知>. Available from: http://www.gov.cn/zwgk/2005-09/08/content_30265.htm

The State Council of the PRC. (2005c). *Several opinions of the state council on speeding up the development of circular economy*. 关于加快发展循环经济的若干意见>. Available from: http://www.gov.cn/zwgk/2005-09/08/content_30305.htm

The State Council of the PRC. (2007). *China's national climate change program*. 中国应对气候变化国家方案>. Available from: http://www.gov.cn/gzdt/2007-06/04/content_635590.htm

The State Council of the PRC. (2011a). *Work plan for greenhouse gas emission control during the 12th Five-Year Plan Period*. "十二五"控制温室气体排放工作方案>. Available from: http://www.gov.cn/zwgk/2012-01/13/content_2043645.htm

The State Council of the PRC. (2011b). *China's policies and actions for addressing climate change*. Available from: http://english.www.gov.cn/archive/white_paper/2014/09/09/content_281474986284685.htm

The State Council of the PRC. (2012). *The 12th Five-Year Plan for energy conservation and emission reduction*. 节能减排"十二五"规划>. Available from: http://www.gov.cn/zwgk/2012-08/21/content_2207867.htm

The State Council of the PRC. (2014). *National new-type urbanization plan (2014–2020)*. 国家新型城镇化规划>. Available from: http://www.gov.cn/zhengce/2014-03/16/content_2640075.htm

The State Council of the PRC. (2016a). *National plan on implementation of the 2030 agenda for sustainable development*. 中国落实2030年可持续发展议程国别方案>. Available from: http://www.gov.cn/xinwen/2016-10/13/5118514/files/44cb945589874551a85d49841b568f18.pdf

The State Council of the PRC. (2016b). *Notice of the state council on issuing the work plan for greenhouse gas emission control during the 13th Five-Year Plan Period*. 十三五"控制温室气体排放工作方案的通知>. Available from: http://www.gov.cn/zhengce/content/2016-11/04/content_5128619.htm

The State Council of the PRC. (2018). *Three year plan of action for winning the war to protect blue skies*. 打赢蓝天保卫战三年行动计划>. Available from: http://www.gov.cn/zhengce/content/2018-07/03/content_5303158.htm

The State Council of the PRC. (2021a). *The guiding opinions on accelerating the establishment and improvement of a green and low carbon circular development economic system*. 关于加快建立健全绿色低碳循环发展经济体系的指导意见>. Available from: http://www.gov.cn/zhengce/content/2021-02/22/content_5588274.htm

The State Council of the PRC. (2021b). *Report on the work of the government*. 政府工作报告>. Available from: http://english.www.gov.cn/premier/news/202103/13/content_WS604b9030c6d0719374afac02.html

Xie, L., Cheshmehzangi, A., Tan-Mullins, M., Flynn, A., & Heath, T. (2020). Urban entrepreneurialism and sustainable development: A comparative analysis of Chinese eco-developments. *Journal of Urban Technology, 27*(1), 3–26.

Zhang, Z. (2010). China in the transition to a low-carbon economy. *Energy Policy, 38*(11), 6638–6653.

Zhao, X. G., Wu, L., & Li, A. (2017). Research on the efficiency of carbon trading market in China. *Renewable and Sustainable Energy Reviews, 79*, 1–8.

Zhou, N., He, G., & Williams, C. (2012). *China's development of low-carbon eco-cities and associated indicator systems* (No. LBNL-5873E). Technical Report, Lawrence Berkeley National Lab (LBNL), Berkeley, CA (United States). https://doi.org/10.2172/1172952

Chapter 8
Low Carbon Transitions: Practices and Lessons

8.1 General Insight of China's Low Carbon Pilot Initiative

The previous chapter described the history and prospects of China's low carbon development. Besides, it discusses China's initiatives at the national or local levels to pursue a low carbon strategy. China has been exploring a low carbon development path and model that suits its reality and successively established three batches of low carbon pilot cities. For example, the carbon dioxide emissions per unit of GDP became a binding target in the 12th Five-Year Plan (FYP) (Liu et al., 2011). Climate change, which was also regarded as a significant issue, was incorporated into China's national economic and social development plans. The goal of controlling GHG emissions and adapting to climate change was served as an important basis for governments at all levels to formulate strategies and plans. The National Development and Reform Commission (NDRC) has put forward three batches of low carbon pilot regions and a group of carbon emission trading pilot areas. Their distribution is shown in Fig. 8.1. China's low carbon policies have also contributed to changes in urban emissions of GHG, such as carbon dioxide. The CO_2 emissions of each city in 2019 are shown in Fig. 8.2.

In this chapter, we review China's low carbon pilot initiative (LCPI) and summarise the experience and practical lessons in China. Since 2008, the Chinese government has continuously deepened the national pilot projects of low carbon provinces, low carbon cities, and promoted pilot projects of low carbon industrial parks, low carbon communities, low carbon cities, and green transportation. Around 2010, regional sustainable development plans and construction projects with "low carbon" and "green" labels emerged in various provinces, cities, and regions. Energy conservation and emission reduction became an important direction of regional sustainable development and construction. In addition, China has promoted low carbon pilot projects and demonstrations in other areas such as green transportation, Carbon Capture, Utilisation and Storage (CCUS), sponge city, etc. For example, the

A. Cheshmehzangi and H. Chen, *China's Sustainability Transitions*,
https://doi.org/10.1007/978-981-16-2621-0_8

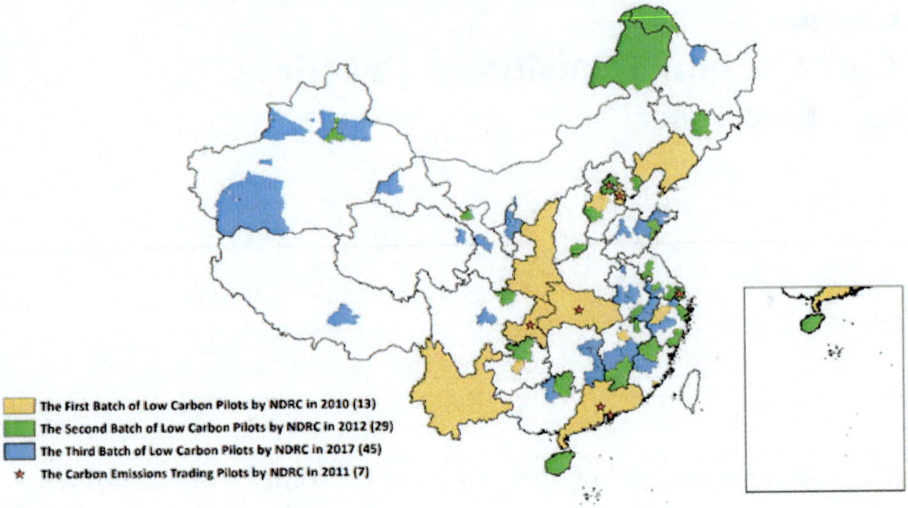

Fig. 8.1 Distribution of pilot cities or provinces in China (Drawn by the Authors, data extracted from Cheng et al., 2019; Khanna et al., 2014)

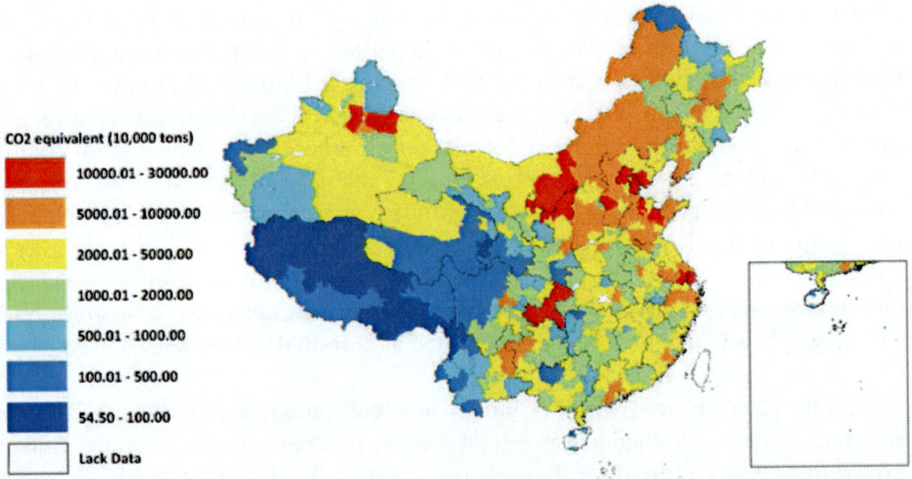

Fig. 8.2 Total carbon dioxide emissions of Chinese cities in 2019 (Drawn by the Authors, data extracted from https://www.ipe.org.cn/index.html)

"Technical Guidelines for Carbon Dioxide Capture, Utilisation and Storage Environmental Risk Assessment (Trial)" (MEE, 2016), which proposed the methods to assess environmental risks of CCUS demonstration projects, was republished in 2016.

8.2 Low Carbon Regional Development: Exploration in Typical Regions

In 2010, "The notice of piloting low carbon provinces and low carbon cities" (NDRC, 2010) confirmed that pilot work would be first carried out in Guangdong, Liaoning, Hubei, Shaanxi, and Yunnan provinces and in eight cities, namely, Tianjin, Chongqing, Shenzhen, Xiamen, Hangzhou, Nanchang, Guiyang, and Baoding. These regions have been asked to introduce low carbon development plans, formulate supporting policies to support low carbon and green development, and accelerate the establishment of low carbon industrial systems (Liu et al., 2012). At the same time, these regions have planned to establish a greenhouse gas emission data statistics and management system and actively advocate low carbon and green lifestyles and consumption patterns. The total annual carbon emissions of energy consumption in Shaanxi, Guangdong, Liaoning, Hubei, and Yunnan provinces increased rapidly. The economic and carbon emissions showed a weak decoupling trend in the 10 years before implementing low carbon policies (Liu et al., 2011). Some studies have shown that the implementation of low carbon pilot policies has significantly reduced the per capita carbon emissions of pilot provinces and cities compared with those before the implementation of low carbon policies and other provinces or cities (Dai & Cao, 2015). LCPI has become extremely important for low carbon development.

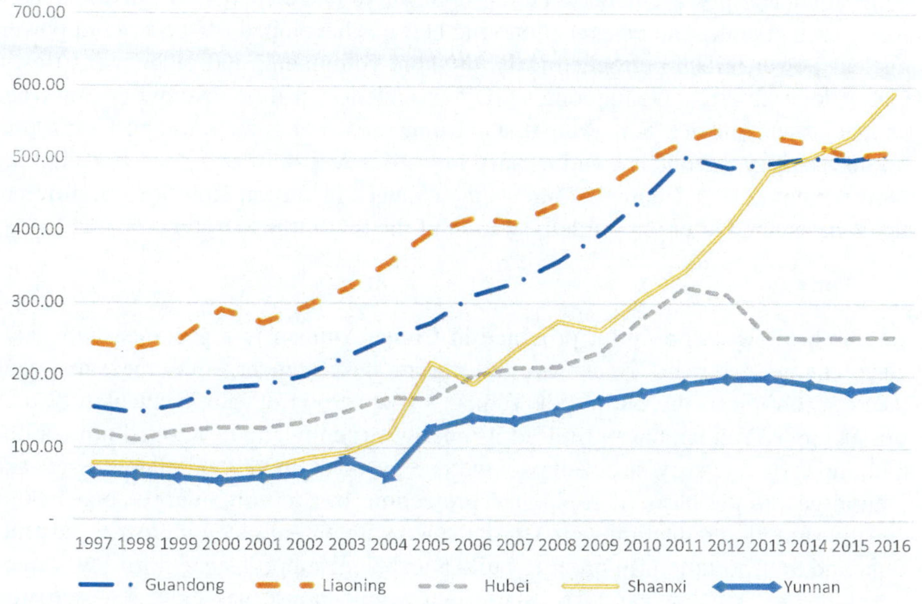

Fig. 8.3 Total apparent CO_2 emissions (mt) of five pilot provinces (Drawn by the Authors, data extracted from https://www.ceads.net/data/province/)

In this part, we select some typical low carbon pilot provinces to review their achievements in low carbon development. Figure 8.3 shows the total carbon emissions of those pilot low carbon provinces in the past period. These five provinces represent typical regions with different economic levels, industrial organisations, and geographical locations in China.

1. **Guangdong**

In 2009, Guangdong's energy consumption per unit of GDP was 0.684 tons of standard coal. Although the total amount of carbon dioxide emissions in Guangdong is increasing, the carbon dioxide emissions per unit of GDP are decreasing year by year. Meanwhile, the carbon productivity is higher than the national level (CMA, 2010). In 2010, Guangdong officially launched the national low carbon pilot program and published the Implementation Plan of Guangdong Province to Carry out the National Low carbon Pilot Province Program. The carbon emissions in Guangdong Province, which has the largest economic scale, population size, and carbon emissions amongst those low carbon pilot provinces, were reduced by about 10% from 2010 to 2015 (Yu et al., 2019). In 2012, Guangdong proposed establishing a carbon emission trading market to make the market mechanism play a role in energy saving, carbon reduction, and total energy consumption control. It is planned that a carbon emission trading mechanism will be basically established by 2020 (Guangdong Government, 2012). Guangdong also promoted the optimisation of the energy mix, large-scale development of nuclear power, increased the proportion of clean energy use, and increased forest carbon sink. The coastal economic belt has developed offshore wind power, nuclear power, green petrochemicals, offshore equipment, and other industries in Guangdong. In 2020, Guangdong's GDP exceeded 11 trillion yuan with an average annual growth rate of 6%, ranking first in China for 32 consecutive years. The proportion of primary, secondary, and tertiary industries was 4.3: 39.2: 56.5 (Guangdong Government, 2021). Figure 8.4 shows the changes in carbon emissions in different fields before and after the implementation of the low carbon strategy.

2. **Yunnan**

As another low carbon pilot province in China, Yunnan is a province with large clean energy resources. To achieve low carbon development, some measures were taken according to the outline of Yunnan's low carbon development plan (2011). For example, Yunnan has optimised its energy structure and developed non-carbon and low carbon energy. It develops wind energy, solar energy, biomass energy, and natural gas on the basis of ecological protection. In addition, energy conservation and consumption reduction were strengthened with emphasis on industry, construction, and transportation. In order to build a technical support system for low carbon development, Yunnan has also established a greenhouse gas emission statistical accounting and management system. Since 2010, Yunnan has undergone 10 years of low carbon reform. By the end of 2019, Yunnan's carbon intensity has decreased by 30.23% compared with 2015. The carbon emission control in the industrial sector has achieved remarkable results. The carbon dioxide emission per unit industrial added

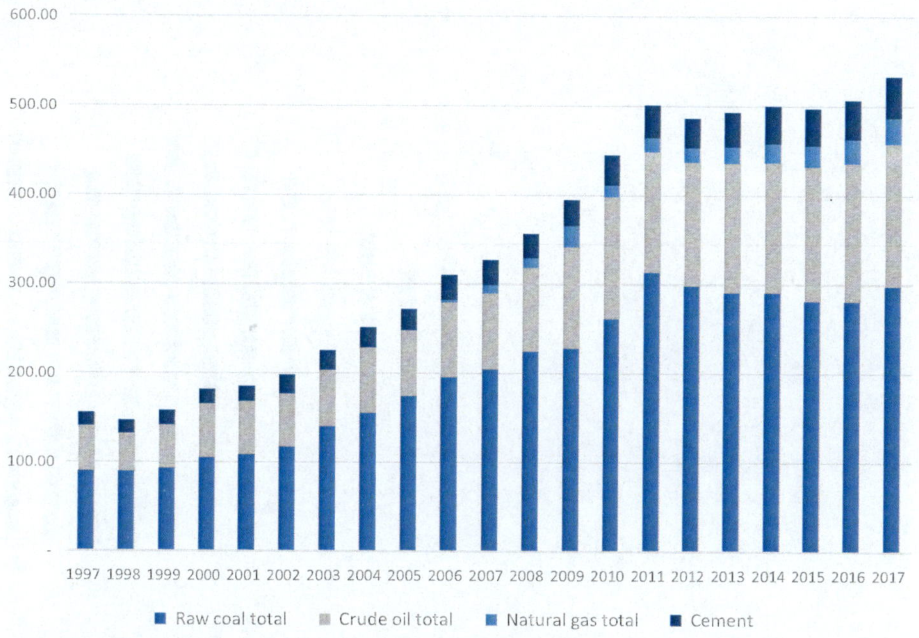

Fig. 8.4 Apparent CO_2 emissions (mt) of Guangdong Province (Drawn by the Authors, data extracted from https://www.ceads.net/data/province/)

value reduced by 28% in 2019 compared with that in 2015 (Yunnan Government, 2020). Figure 8.5 shows the changes in carbon emissions in different fields before and after the implementation of the low carbon strategy in Yunnan.

3. **Liaoning**

Liaoning Province is a traditional heavy industry base in China, and industrial carbon dioxide emissions (industrial energy + industrial process) account for a large proportion in most cities. Liaoning Province 2014–2015 energy conservation and emission reduction low carbon development action plan (Liaoning Provincial Development and Reform Commission, 2014) proposed mainly measures such as follows. Liaoning has developed low energy consumption and low emission industries and strengthened the restrictive role of energy assessment and environmental impact assessment. While accelerating the construction of energy conservation, emission reduction, and carbon reduction projects, Liaoning promoted energy conservation and emission reduction in key sectors such as industry, construction, transportation, and public institutions. Besides, the Liaoning government has promoted green finance, market-based energy conservation, and emission reduction mechanisms implemented energy efficiency labeling and certification for energy-saving and low carbon products. Figure 8.6 shows the changes in carbon emissions in different fields before and after the implementation of the low carbon strategy in Liaoning.

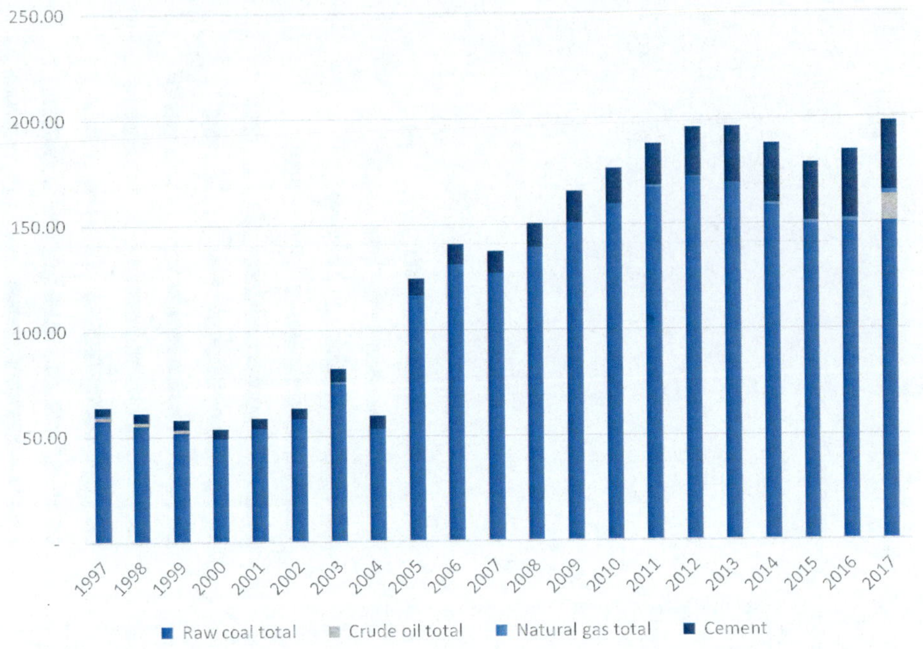

Fig. 8.5 Apparent CO_2 emissions (mt) of Yunnan Province (Drawn by the Authors, data extracted from https://www.ceads.net/data/province/)

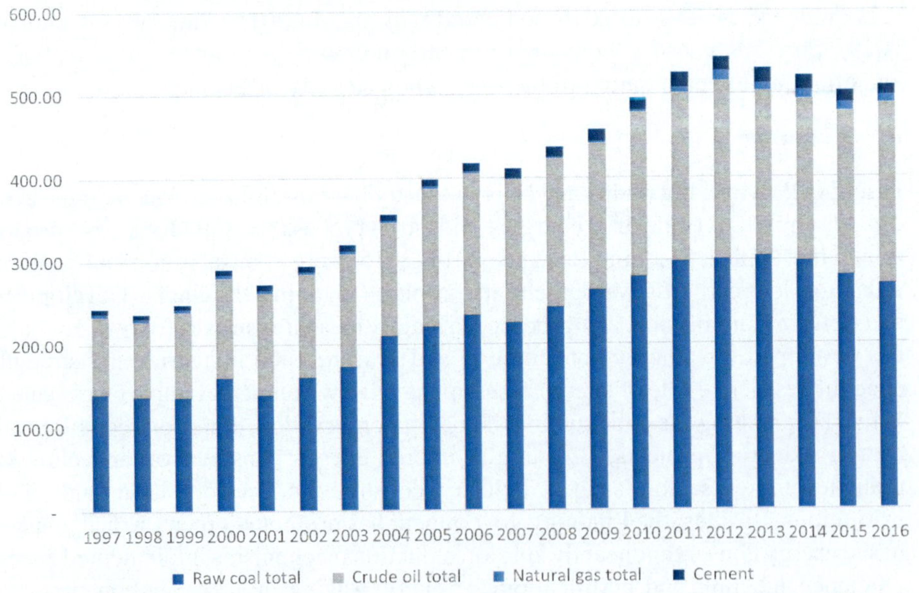

Fig. 8.6 Apparent CO_2 emissions (mt) of Liaoning Province (Drawn by the Authors, data extracted from https://www.ceads.net/data/province/)

8.3 Pilot Low Carbon Cities and Typical Cases

At the beginning of 2008, Shanghai and Baoding became the first cities to join a new World Wildlife Fund (WWF) initiative to explore low carbon development strategies for China's urban areas. To explore the effective mode of urban low carbon development and promote and apply the successful experience, China identified Tianjin, Chongqing, and other cities as the first batch of low carbon pilot cities in 2010 (NDRC, 2010). In 2012, Beijing, Shanghai, and other cities were selected as the second batch of national low carbon pilot cities (NDRC, 2012). Wuhai, Inner Mongolia Autonomous Region, and other cities were selected as the third batch of pilot cities (NDRC, 2017). Low carbon pilot cities mainly achieve carbon intensity reduction by improving energy efficiency and the upgrading of industrial structure. The effect of industrial upgrading shows a trend of increasing year by year (Zhou et al., 2019). Carbon emissions and energy-intensive activities in the pilot cities have dropped significantly. Hangzhou, Xiamen, and Shenzhen have reduced carbon emissions by more than 200,000 tons per year due to low carbon transport and construction projects. From 2010 to 2011, the reduction of carbon dioxide emissions per unit of GDP in pilot cities was 88.9% larger than that of other cities in the provinces (Cheng et al., 2019). The average level of CO_2 emission per unit of GDP in pilot low carbon cities increased gradually from the east to the west (Song et al., 2015). With the expansion of the urban permanent population, CO_2 emissions per unit of GDP in the pilot cities gradually decreased. The per capita CO_2 emissions in the large cities were the lowest (as shown in Table 8.1). At the same time, by comparing with similar areas, the results of low carbon work and carbon reduction targets of pilot cities are generally better than those of similar areas.

In the past 10 years, China's central government and local governments have taken corresponding policies or actions to reduce carbon emissions in industry, energy, construction, and transportation (as shown in Fig. 8.7) (Cheng et al., 2019). These pilot cities and provinces were selected based on geographic, social, and economic diversity and representativeness which was asked to develop and propose a low carbon development plan, formulate supporting policies, develop a low carbon industry, establish CO_2 emission statistics and data management system and encourage low carbon lifestyles and consumption (Khanna et al., 2014). In this part, we select some typical cases to analyse the results of their low carbon actions.

Table 8.1 CO_2 emission of low carbon pilot cities in different urban sizes (Song et al., 2015)

City type	Small	Medium	Large	Super	Great
CO_2 emission per unit of GDP (t CO_2/ten thousand yuan)	4.64	3.81	2.88	2.46	1.88
CO_2 emissions per capita (t CO_2/per)	23.12	17.18	10.21	11.89	13.04
Average resident population (ten thousand)	46.59	59.80	216.54	591.12	1632.60

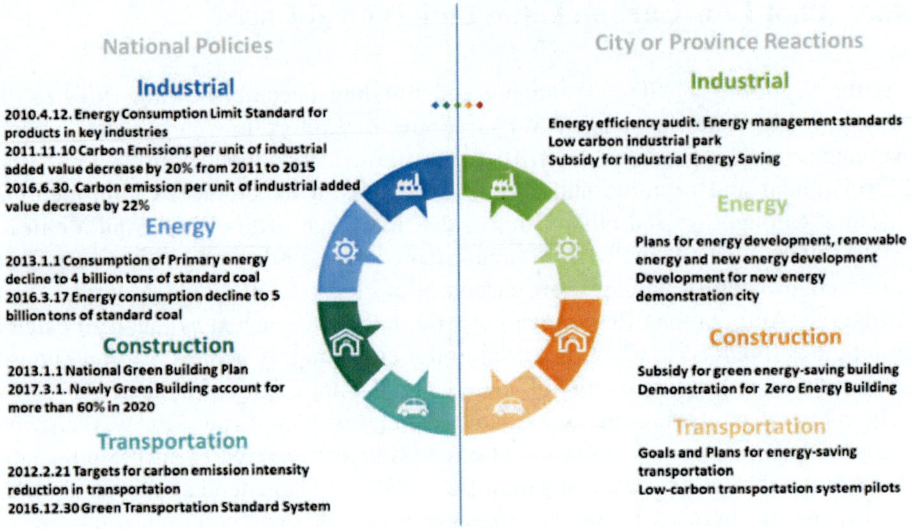

Fig. 8.7 Policies developed by Chinese central and local government (Drawn by the Authors, data extracted from Cheng et al., 2019)

1. **Shanghai**

Shanghai is the largest city in China and the center of financial and technological innovation in China, which has a population of more than 23 million. In March 2011, Shanghai launched the construction of the first batch of low carbon development practice areas in eight areas, including Hongqiao Business District and Chongming County (Shanghai Municipal Development and Reform Commission, 2011). At the same time, a complete evaluation system has been established for the practice area. In May 2017, Shanghai selected the Shanghai World Expo Site, Shanghai International Tourism Resort, and other areas to carry out the construction of the second batch of low carbon development practice areas (Shanghai Municipal Development and Reform Commission, 2017). Low carbon initiatives in those practice areas are shown in Table 8.2. Shanghai proposes to reach a carbon peak by 2025. Shanghai's energy structure has been continuously optimised, and energy intensity has been gradually reduced. Since the 12th FYP, Shanghai's greenhouse gas emission intensity has dropped by 6% annually, thanks to industrial restructuring and technological upgrading (Tanjiaoyi, 2019). Shanghai plans to increase the share of public transport in the city's total traffic to 50% by 2035. Shanghai's citizens will use electric cars on a massive scale. By 2050, the share of electric passenger vehicles will rise to 68%.

2. **Shenzhen**

As one of the earliest low carbon pioneers in China, Shenzhen has carried out several low carbon and eco projects in many areas (Cheshmehzangi et al., 2018a, 2018b). The city is well known for its major initiatives on progressive low carbon development strategies, including industrial restructuring, conversion of high-polluting industries

Table 8.2 Low carbon initiatives in Shanghai Low carbon Practice Area

Target	Actions
Urban regeneration	• Transformation of old industrial areas
Low carbon energy	• Clean and efficient diversified energy supply system • Regional natural gas distributed energy supply system
Low carbon building	• Promote the application of green buildings and building information modelling (BIM) technology • Building energy consumption monitoring and management platform
Low carbon transportation	• Promote the mode of shared travel • Demonstration of green and low carbon transport hub • Intelligent traffic management system • TOD planning technology model, the construction of BRT and slow traffic systems
Resource utilisation	• Sponge city construction • Domestic waste classification • Solid waste recycling and disposal facilities
Green infrastructure	• Drainage system modification • Green garden and space
Low carbon culture	• Low carbon travel publicity • Low carbon office advocacy • Low carbon life publicity

to low-polluting industries, diversified revenue generation, extensive investment in smart finance and trade, the rapid development of infrastructure for electric vehicles, and achieving 100% electric taxis and public transport vehicles. The city has adapted district-level strategies that are first developed in its core districts and then replicated in other districts. Over the years, Shenzhen has developed into a smart city role model (Tan-Mullins et al., 2017) and a successful example of a progressive low carbon city with citywide initiatives and strategies.

Shenzhen is one of the first low carbon pilot cities and carbon emission trading pilot cities. Although its economic aggregate has maintained a relatively high growth rate, the growth rate of energy consumption and carbon emissions has been declining year after year. At the same time, Shenzhen has also carried out various energy-saving and emission reduction actions, such as clean production, green building, energy conservation renovation, energy contract management, etc. Shenzhen's energy consumption and carbon emission intensity per unit of GDP have dropped to one-third and one-fifth of the national average, respectively. In 2020, the proportion of emerging industries in Shenzhen reached 37% of the total industries. Shenzhen's installed capacity of clean power, including nuclear power and gas power, accounts for 77% of its total installed capacity (Shenzhen Government, 2021). In order to achieve the peak of carbon emissions, Shenzhen has taken emission reduction measures from different aspects. These include five key areas, with specific targets on low carbon development. The first aspect is clean and low carbon energy.

The targets include control of coal consumption, increase in the installed capacity of clean energy power generation and renewable energy capacity development. The second aspect is reducing carbon emissions in industry, including the improvement of energy efficiency in industry, optimisation and/or upgrading of the manufacturing sector, and building green data centers across the city. The third aspect is focused on the development of green and low carbon transportation systems. In this target group, the city focuses on promoting new energy vehicles and supporting facilities such as charging and hydrogenation built and integrated into the city's existing infrastructure. The fourth aspect is to build green energy saving, specifically to expand on the application scale of prefabricated buildings and transforming existing public and residential buildings. Lastly, the fifth aspect is centered around the overarching carbon emission trading scheme, which includes the construction of near-zero carbon emission demonstration projects and reforms the mechanisms for climate investment and financing.

3. **Beijing**

Beijing has been carrying out air pollution control, promoting the optimisation of industrial structure and the transformation of clean energy. In 2020, Beijing's carbon intensity was more than 23% lower than that of 2015 (XinhuaNet, 2021). In addition, Beijing also released the "Technical guidelines for low carbon community assessment" (Government of Beijing, 2016) to launch the practice of low carbon community. During the 13th FYP period, Beijing has realised the optimisation and transformation of its energy structure, which is dominated by electricity, natural gas, and oil products. In 2018, there were basically no coal-fired boilers in the city. Beijing's coal consumption had fallen from a peak of more than 30 million tons to 1.73 million tons in 2020. The share of clean and quality energy such as electricity and gas increased to 98.1%. In 2013, as one of the first provinces and cities in China to carry out carbon emission trading pilot projects, Beijing officially launched the carbon market. In 2020, the quota trading volume of Beijing's pilot carbon market reached 4.7 million tons, and the trading volume reached 245 million yuan (GMW, 2021).

4. **Guiyang**

Since Guiyang was rated as one of the cases of the first batch of low carbon cities in China in 2010, it has coordinated the urban economic and social development with ecological civilization construction while exploring the low carbon road of "development in protection" in the west of China, to realise economic growth without sacrificing environmental quality (Cecelia & Tian, 2015). In addition, Guiyang has developed and promoted gas electric dual fuel hybrid electric vehicle and methanol fuel vehicle, which has produced good emission reduction effect (Tian et al., 2015). Guiyang takes ecological civilization as the general strategy of urban development. It vigorously promotes the service industries, improves resource-based industries with its ecological, climate, location, and cultural advantages. It aims to explore low carbon development mode and find a low carbon development path for economically underdeveloped cities (Zheng, 2019), which rely on renewable resources.

5. **Suzhou**

Suzhou Development and Reform Commission organised the compilation of "Suzhou Low carbon Development Plan", which elaborated the guiding ideology, development goals, main tasks, and key lines of the city's low carbon development work by 2020 (Suzhou Government, 2014). Suzhou has optimised the economic structure, promoted energy conservation and emission reduction, adjusted the energy structure and energy efficiency, increased forest carbon sequestration and other policies, further improved the system, overall planning, clear objectives, and implemented tasks, and achieved remarkable results in comprehensively promoting the construction of low carbon development (Gu, 2016). Since 2005, the proportion of carbon sink and carbon source in Suzhou has increased year by year, and the carbon sink had gradually increased from 947,200 tons in 2005 to 3.015 million tons in 2017 (Liu et al., 2019).

6. **Wuhan**

Wuhan puts forward the concept of integrating top-level design into low carbon development. Wuhan City Master Plan (2010–2020) has made valuable exploration in coordinating ecological protection and urban development. Many low carbon research results were used to guide the government to implement policies. In terms of adjusting the energy structure, Wuhan introduced a comprehensive improvement project of energy conservation and environmental protection for coal-fired boilers. In general, based on the practice of low carbon development in Wuhan, the experimental model of low carbon reform in Wuhan is summarised as the development model of a comprehensive "low carbon society". Through the integration and transformation development of new urbanisation, the mode energy conservation and improves energy efficiency, comprehensively saves energy, and reduces carbon in industry, construction, transportation, and urban carbon sequestration (Wu & Wan, 2020). It guides low carbon production and lifestyle, drives the reduction of energy consumption and carbon emissions of the whole society.

8.4 Typical Low Carbon Communities and Projects

In March 2014, the pilot construction of low carbon communities in China was officially launched (NDRC, 2014). To summarise, low carbon community pilot cases are divided into three types: (1) new urban community, (2) existing urban community, and (3) rural community pilot. Among all kinds of low carbon community pilot projects, existing urban community pilot projects have the widest scope and the largest number (Fu et al., 2020). In recent years, pilot communities have carried out low carbon management, construction, publicity, and demonstration activities with their own characteristics. Their main practices are presented in Table 8.3.

In order to promote the pilot construction of low carbon communities, the "Guidelines for the Construction of Pilot Low carbon Communities" (NDRC, 2015) was

Table 8.3 Major initiatives in China's low carbon communities (The translated data from Fu et al., 2020)

Dimensions	Main measures
Leading concept	• Formulate the Community Low carbon Development Plan • Integrate low carbon elements into existing community planning • Make low carbon community work plan • Put forward the construction goals • Carry out low carbon transformation and hold competitions
Culture and lifestyle	• Evaluation of low carbon households and low carbon enterprises • Low carbon life guidance and manuals • Promote low carbon products • Advocate low carbon travel • Low carbon publicity and education • Recycling activities
Management	• Community management system and regular assessment • Barter exchange network platform and other online platforms • Data and fine management • Household carbon emission statistics survey and annual GHG inventory
Renovation of buildings	• Transformation of building exterior wall, shading energy saving • Promote the contract energy management model • Energy-saving renovation of public areas • Photovoltaic power generation, surplus electricity online, promote solar heating equipment and other photovoltaic products
Infrastructure	• Reasonable planning of community layout • Construction of low carbon transportation facilities and installation of coin-operated electric bicycle charging piles • Renovation of water supply and drainage facilities and the establishment of rainwater collection system • Garbage/rubbish disposal facilities • Improve the efficiency of community heating
Livable environment	• Improve natural ecosystems • Implement ecological and environmental planning • Build low carbon public places

released in 2015. It further clarified the basic requirements, pilot implementation, and safeguard measures in the pilot process of low carbon communities. In October 2016, the State Council issued the "Notice of the State Council on Issuing the Work Plan for Greenhouse Gas Emission Control during the 13th Five-Year Plan Period" (The State Council, 2016). This notice proposed pilot demonstration of low carbon development in innovative regions, the implementation of about 1,000 pilot low carbon communities, and the organisation of 100 national low carbon demonstration communities.

For example, Yanji No.7 Village in Yangpu, Shanghai, has actively carried out low carbon community construction. It increased the promotion and application of energy-saving and low carbon technology and products to strengthen the publicity of the low carbon concept (Liu et al., 2015).

Wuhan Baibuting community, which is the largest residential project in Wuhan, advocates the self-management of residents. This community also has been adhering to the concept of green ecology. Baibuting community adopted means to achieve low carbon goals, such as mixed land resources and reduction of the transportation cost and energy consumption (Min & Xiong,). Changxindian low carbon community in Beijing formulated a concept plan to establish an environmentally friendly and resource-efficient community (ARUP, 2010). These communities have achieved remarkable results in environmental protection, energy-saving equipment, and smart communities. They pay attention to the disposal of garbage and rainwater recycling, energy-saving lamp source transformation. In addition, they also proposed a change in the way of water supply and the use of Internet technology and smart devices to improve management efficiency.

8.4.1 China's Carbon Emissions Trading Markets

From 2011 to 2020, carbon markets in China's pilot provinces and cities cover more than 20 industries, such as steel, electricity, and cement, involving nearly 3,000 enterprises, with a cumulative turnover of more than 400 million tons and a total turnover of more than 9 billion yuan (Economic Daily, 2020). There are mainly two types of trading in China's carbon emission trading market, namely cap and quota trading and project emission reduction trading. The trading object of the former is mainly the carbon emission quota allocated by emission control enterprises. In contrast, the trading object of the latter is mainly the emission reduction certificate obtained through the implementation of projects to reduce greenhouse gases. "Chinese Certified Emission Reduction (CCER)" is the voluntary emission reduction recorded by the ecological and environmental authorities. Carbon trading is applied as a market mechanism for reducing China's carbon dioxide emissions and mitigating climate change. It started with the first official document, "Decision of the State Council on Accelerating the Fostering and Development of Strategic Emerging Industries", in 2010 (Wen & Xu, 2018). In 2011, the Chinese government issued the "Notice on Carrying out the Carbon Emissions rights Trading Pilot Work" (NDRC, 2011), which approved the implementation of pilot carbon emission trading programs in Beijing, Tianjin, Shanghai, Chongqing, Hubei, Guangdong, and Shenzhen. The "Interim Measures for the Administration of Carbon Emission Permit Trading" (NDRC, 2014) was released in December 2014. The Chinese government proposed to carry out emission quota and national certification of voluntary emission reduction trading activities in accordance with this document. "Measures for the Administration of Carbon Emission Trading (for Trial Implementation)",approved by the MEE and came into force on February 2021 (MEE, 2021). Trading of carbon emission rights is conducted through the National Carbon Emission Trading System and may take the form of agreement transfer, one-way bidding, or other conforming methods. Before the establishment of the national carbon emission trading system, the pilot areas had explored and established a variety of trading methods, including overall bidding, partial bidding, pricing, listing and selection, auction trading, and so on. Table 8.4 shows the changes in China's carbon trading market policies.

8.5 Progresses, Limitations, and Prospects

The construction of low carbon cities has become the focal point of China's efforts to meet its climate change goals and promote the transition of economic development to green and low carbon development. In a way, the LCPI links carbon emissions to economic development. In this regard, we can argue that China's low carbon cities are exploring new development models in energy, transportation, construction, industry, and residential life (Xie et al., 2020). China's low carbon pilot cities show different emission reduction characteristics due to different conditions. Therefore, urban low carbon transformation should be based on the stage of urban economic development and resources to form a characteristic spatial pattern, industrial structure, mode of production, and way of life. China's efforts to promote low carbon urban development actually force cities to achieve low carbon transition faster from carbon emission targets. The pilot low carbon cities in China have initially formed the policy formulation from the top-level strategy at the national macro-level to the urban medium level.

China's experimental projects and practices to date are robust examples of low carbon transitions and achieving urban sustainability targets, specifically on greener development and towards the reductions of GHG emissions, carbon emissions, and energy use. China has and continue to experience three stages of gathering and learning from lessons, demonstration of experimental projects, and exporting the findings and achievements to larger scales (i.e., through scaling up), replicating the success examples, and promoting them through policies and practices (Cheshmehzangi et al., 2018a). In the future, China needs to further promote the low carbon transformation of cities through market-based tools, such as carbon trading and financing means such as green finance. In this chapter, some of these are explained, but we also highlight existing gaps and limitations with tangible scope for further development. For example, China's eastern regions have the largest population and the country's most dynamic economy. Therefore, those regions are expected to form a coordinated regional low carbon emission reduction strategy. The central and western regions can promote cooperation between cities through the allocation of regional production factors. More importantly, the experience gained from China's pilot low carbon cities can be used as a reference for cities in other developing countries.

To follow up, in the next three chapters, we explore some of the key global lessons on low carbon development, explore China's more recent progress towards carbon neutrality, and highlight some key directions for the next steps. In these forthcoming chapters, we delve into the context of China and low carbon strategies, lessons, and potential paradigms, before concluding the book with further discussions, viewpoints, and suggestions.

Table 8.4 China's carbon emission trading policies

Law name	Effective time	Enacting agency	Legislative level	Main content
Notice on Carrying out the Carbon Emissions rights Trading Pilot Work	29-Oct-11	National Development and Reform Commission	Normative documents	Approval of pilot projects for carbon emissions trading in seven provinces and cities: Beijing, Tianjin, Shanghai, Chongqing, Hubei, Guangdong, and Shenzhen
Interim Measures for the Administration of Carbon Emissions Trading	10-Jan-15	National Development and Reform Commission	Departmental rules	The basic framework of China's unified carbon emissions trading market is clarified, and a secondary management system including national and provincial authorities will be established
Interim Regulations on the Management of Carbon Emissions Trading (Draft for Solicitation of Comments)	29-Mar-19	Ministry of Ecology and Environment	Departmental rules	It is planned to formulate administrative regulations to form a higher-level law in the field of carbon emissions trading, which will serve as the legal basis for the implementation of national carbon emissions trading and provide for the core issues of carbon emissions trading. The regulations have not yet been promulgated and entered into force
National Carbon Emissions Trading Management Measures (for Trial Implementation) (Draft for Solicitation of Comments)	28-Oct-20	Ministry of Ecology and Environment	Departmental rules	Provision of principles for the registration, trading, and settlement of carbon emission rights across the country

(continued)

Table 8.4 (continued)

Law name	Effective time	Enacting agency	Legislative level	Main content
National Carbon Emission Rights Registration and Transaction Settlement Management Measures (for Trial Implementation) (Draft for Solicitation of Comments)	28-Oct-20	Ministry of Ecology and Environment	Departmental rules	Make specific provisions on the relevant elements of the national carbon emission rights registration, trading, and settlement activities
Measures for the Administration of Carbon Emission Trading (for Trial Implementation)	1-Feb-21	Ministry of Ecology and Environment	Departmental rules	Organise the establishment of a national carbon emission rights registration institution and a national carbon emission rights trading institution, and organise the construction of a national carbon emission rights registration and registration system and a national carbon emission rights trading system. The National Carbon Emissions Trading Agency is responsible for organising and carrying out centralised and unified trading of national carbon emission rights

References

ARUP. (2010). Beijing Changxindian low carbon community concept plan. *Urbanism and Architecture, 02*, 44–46.

Cecelia & Tian, D. (2015). Orientation of low-carbon pilot cities in the next five years—Review of the process of the pilot work of low-carbon cities in Guiyang and Zunyi. *World Environments, 6*, 68–71. Available from http://www.wem.org.cn/zzlm/rdgz/201602/W020180903671713158843.pdf

Cheng, J. H., Yi, J. H., Dai, S., & Xiong, Y. (2019). Can low-carbon city construction facilitate green growth? Evidence from China's pilot low-carbon city initiative. *Journal of Cleaner Production.*

Cheshmehzangi, A., Xie, L., & Tan-Mullins, M. (2018a). The role of international actors in low-carbon transitions of Shenzhen's International Low Carbon City in China. *Cities, 74*, 64–74.

Cheshmehzangi, A., Li, H., & Yang, X. (2018b). Low Carbon and Climate Resilient urban development in China. ADB Report, developed for the National Development Reform Committee (NDRC), Beijing, China, Oct 2018.

China Meteorological Administration (CMA). (2010). *Guangdong Province has officially launched the national pilot project of low carbon province.* Available from: http://2011.cma.gov.cn/qhbh/newsbobao/201011/t20101103_80799.html

Dai, R., & Cao, J. H. (2015). The effect of China's first low-carbon pilot—DID estimation of five cities and eight provinces. *Science and Technology Management Research, 12*, 56–61.

Economic Daily. (2020). *The cumulative turnover of the pilot carbon emission trading program exceeded 9 billion yuan.* Available from http://www.gov.cn/xinwen/2020-11/03/content_5556838.htm

Fu, L., Yang, X., & Di, Z. (2020). Practice, experience, challenges and suggestions for low carbon community pilots in China. *Environmental Protection, 48*; *693*(22), 64–68.

GMW.cn. (2021). *Beijing: Carbon intensity in 2020 is expected to fall by more than 23% from 2015 levels.* Available from https://m.gmw.cn/baijia/2021-01/21/1302061123.html

Gu, H. (2016). Suzhou: unswervingly take the road of green cycle and low carbon development. *China Economic Trade Herald, 19*, 54–55.

Khanna, N., Fridley, D., & Hong, L. (2014). China's pilot low-carbon city initiative: a comparative assessment of national goals and local plans. *Sustainable Cities & Society., 12*, 110–121.

Liaoning Provincial Development and Reform Commission. (2014). *Liaoning Province 2014–2015 energy conservation and emission reduction low carbon development action plan* <辽宁省2014–2015年节能减排低碳发展行动计划>. Available from http://www.ln.gov.cn/zwgkx/lnsrmzfgb/2014/d20q_103361/gwywj/201412/t20141222_1510017.html

Liu, J., Wang, R., Sun, Y. W., Shu, S. Y., & Xiao, L. S. (2012). Analysis of development path for low-carbon pilot provinces in China. *China Population, Resources and Environment, 3*, 56–62.

Liu, Z., Geng, Y., Xue, B., Fu, J. F., & Tang, X. F. (2011). Relationships between economic growth and CO_2 emissions for low-carbon pilot provinces in China. *Resources Science, 33*(4), 620–625.

Liu, C., Chen, M., & Liu, H. (2015). Combined with reality through diversified measures to promote low carbon community construction work Yanji No. 7 community low carbon community construction gets preliminary results in Yangpu District. *Shanghai Energy Conservation, 11*, 598–599.

Liu, H., Zhang, Z., & Yuan, S. (2019). Energy consumption status and carbon emission reduction path of Suzhou Industrial Enterprises. *Northern Economy and Trade, 5*, 122–125.

Min, L., & Xiong, B. (2012). Livable neighborhood planning strategies: wuhan case. *Planners, 28*(6), 18–23.

Ministry of Ecology and Environment of the PRC (MEE). (2016). *Technical guidelines for carbon dioxide capture, utilization and storage environmental risk assessment (Trial)* <二氧化碳捕集,利用与封存环境风险评估技术指南 (试行))>. Available from http://www.mee.gov.cn/gkml/hbb/bgt/201606/t20160624_356016.htm

Ministry of Ecology and Environment of the PRC (MEE). (2021). *Measures for the administration of carbon emission trading (for trial implementation)* <碳排放权交易管理办法(试行)>. Available from http://www.gov.cn/zhengce/zhengceku/2021-01/06/content_5577360.htm

National Development and Reform Commission (NDRC). (2010). *Notice on conducting pilot work of low-carbon provinces and low-carbon cities* <关于开展低碳省区和低碳城市试点工作的通知>.Available from http://www.gov.cn/zwgk/2010-08/10/content_1675733.htm

National Development and Reform Commission (NDRC). (2011). *Notice on Carrying out the Carbon Emissions rights Trading Pilot Work* <关于开展碳排放权交易试点工作的通知>. Available from: https://zfxxgk.ndrc.gov.cn/web/iteminfo.jsp?id=1349

National Development and Reform Commission (NDRC). (2012). *Notice on launching the second batch of national low carbon provinces and low carbon cities pilot work* <关于开展第二批国家低碳省区和低碳城市试点工作的通知>. Available from: http://fgw.czs.gov.cn/fzggdt/dqjjylxsh/content_279370.html

National Development and Reform Commission (NDRC). (2014). *Interim measures for the administration of carbon emission permit trading* <碳排放权交易管理暂行办法>. Available from http://www.gov.cn/gongbao/content/2015/content_2818456.htm

National Development and Reform Commission (NDRC). (2014). *Notice on conducting pilot work of low-carbon communities* <关于开展低碳社区试点工作的通知>. Available from http://www.gov.cn/xinwen/2014-03/27/content_2648003.htm

National Development and Reform Commission (NDRC). (2015). *Guidelines for the pilot construction of low-carbon communities* <低碳社区试点建设指南>. Available from https://www.ndrc.gov.cn/hdjl/yjzq/201411/W020190927558591660768.pdf

National Development and Reform Commission (NDRC). (2017). *Notice on launching the third batch of national low-carbon city pilot projects* <关于开展第三批国家低碳城市试点工作的通知>. Available from http://www.gov.cn/xinwen/2017-01/24/content_5162933.htm

Shanghai Municipal Development and Reform Commission. (2011). *Notice on the pilot work of low-carbon development practice area in eight areas including Hongqiao Business District* <关于在虹桥商务区等8个区域开展低碳发展实践区试点工作的通知>. Available from https://fgw.sh.gov.cn/zgjjl/20110322/0025-16601.html

Shanghai Municipal Development and Reform Commission. (2017). *Notice on the implementation of the second batch of low-carbon community pilot construction work* <关于开展本市第二批低碳社区试点创建工作的通知>. Available from https://fgw.sh.gov.cn/cxxxgk/20170527/0025-27581.html

Song, Q. J., Wang, Y. F., & Qi, Y. (2015). Study on present status of carbon emissions in China's low-carbon pilot cities. *China Population, Resources and Environment, 1*, 78–82.

Tanjiaoyi. (2019). *Discussion on low-carbon path in the 14th Five-Year Plan of Shanghai*. Available from http://www.tanjiaoyi.com/article-28405-1.html

Tan-Mullins, M., Cheshmehzangi, A., Chien, S., & Xie, L. (2017). *Smart-eco cities in China: Trends and city profiles 2016*. Exeter: University of Exeter (SMART-ECO Project).

The People's Government of Beijing Municipality. (2016). *Technical guidelines for low-carbon community assessment* <低碳社区评价技术导则>. Available from http://sthjj.beijing.gov.cn/bjhrb/resource/cms/article/679767/10930477/2021012816215335743.pdf

The People's Government of Guangdong Province. (2012). *Implementation plan of carbon emission trading pilot work in Guangdong Province* <广东省碳排放权交易试点工作实施方案>. Available from http://www.gd.gov.cn/gkmlpt/content/0/141/post_141049.html#7

The People's Government of Guangdong Province. (2021). *Yunnan has exceeded its carbon intensity reduction target for the 13th Five-Year Plan ahead of schedule* <云南提前超额完成"十三五"碳强度下降目标>. Available from http://www.yn.gov.cn/ywdt/bmdt/202007/t20200703_206946.html

The People's Government of Shenzhen Municipality. (2021). *Shenzhen plans to adopt five major measures to reduce carbon emissions* <深圳拟推5大举措降低碳排放>. Available from http://www.sz.gov.cn/cn/ydmh/zwdt/content/post_8586528.html

The People's Government of Suzhou Municipal. (2014). *Suzhou low carbon development plan* <苏州市低碳发展规划>. Available from http://www.suzhou.gov.cn/szsrmzf/zfwj/201403/4beaff22ae7d465a830936b1d2bdaf5f.shtml

The People's Government of Yunnan Province. (2020). *Report on the work of the Guangdong government* <广东省政府工作报告>. Available from http://www.gd.gov.cn/gkmlpt/content/3/3185/post_3185894.html#45

The State Council of the PRC. (2016). *Notice on issuing the work plan for Greenhouse Gas Emission Control during the 13th five-year plan period* <"十三五"控制温室气体排放工作方案的通知>. Available from http://www.gov.cn/zhengce/content/2016-11/04/content_5128619.htm

Tian, D. Y., Liu, Q., Tian, C., & Zheng, X. Q. (2015). Progress and suggestions of guizhou low-carbon pilot city work. *China Economic & Trade Herald, 35*, 6–7.

Weng, Q., & Xu, H. (2018). A review of China's carbon trading market. *Renewable and Sustainable Energy Reviews, 91*, 613–619.

Wu, X., & Wan, Y. (2020). Analysis on the path and mode of Wuhan low carbon city pilot construction. *Industrial Safety and Environmental Protection, 46*(2), 102–106.

Xie, L., Cheshmehzangi, A., Tan-Mullins, M., Flynn, A., & Heath, T. (2020). Urban entrepreneurialism and sustainable development: A comparative analysis of Chinese Eco-Developments. *Journal of Urban Technology, 27*(1), 3–26.

XinhuaNet. (2021). *Beijing has embarked on a new path of green and low-carbon development*. Available from http://www.xinhuanet.com/energy/2021-03/17/c_1127219888.htm

Yu, X., Shen, M. H., Wang, D., & Tadala, B. (2019). Does the low-carbon pilot initiative reduce carbon emissions? Evidence from the application of the synthetic control method in Guangdong Province. *Sustainability, 11*(14), 3979.

Yunnan Provincial Development and Reform Commission. (2011). *Outline of Yunnan low carbon development plan* <云南省低碳发展规划纲要>. Available from http://yn.gov.cn/zwgk/zfgb/17yqgb/201904/P020190708861334344340.pdf

Zhen, W. (2019). Opportunities and challenges of green and low carbon development in Guiyang. *Energy Conservation, 38*(5), 119–121.

Zhou, D., Zhou, F., & Wang, X. (2019). Impact of low-carbon pilot policy on the performance of urban carbon emissions and its mechanism. *Resources Science, 3*, 546–556.

Chapter 9
Learning from Main Low Carbon Strategies

9.1 Learning from Global Models: Carbon Reduction Measures

In this chapter, we highlight some of the key low carbon strategies. To start with, we emphasise the role of carbon reduction measures in achieving low carbon development. We urge relevant stakeholders to consider energy policy and regulations that suggest carbon reduction practices, such as buildings, communities, cities, etc. The examples of life-cycle assessment (LCA) and reduction of embodied energy (Cheshmehzangi & Butters, 2017) are good examples of transdisciplinary and multi-objective methods to carbon reductions. The nexus between carbon reduction and sustainability enhancement is an area that requires further attention, especially in the studies of the built environments. This relationship is studied well in residential areas (Ismailos & Touchie, 2017; Laes et al., 2018), as well as in studies that relate to building energy consumption reductions (Lin et al., 2017), renewable energies (Bagheri et al., 2019; Yu et al., 2020), and general carbon reduction pathways and progresses (Druckman et al., 2008). Throughout this chapter, we aim to highlight the key implications and lessons that could lead towards carbon reduction opportunities and gradual progression of achieving carbon neutrality.

In achieving sustainable urbanisation, carbon emission reductions and energy-related factors are considered vital research areas (Cheshmehzangi, 2020a; b). In particular, the focus on carbon intensity reduction (Cheng & Yao, 2021), energy efficiency enhancement (Zhang, 2019), and lifecycle carbon footprint reduction (Kairies-Alvarado et al., 2021), suggest ways of dealing with climate issues (Gil & Bernado, 2020). The specific examples related to low carbon measures include energy-saving approaches (Li & Yao, 2020) and considering the built environments' energy performances (Dawodu & Cheshmehzangi, 2017; Deng & Cheshmehzangi, 2018; Deng et al., 2020; Ding et al., 2020; Shang et al., 2020). Earlier examples look into successful community-based practical methods (Kellett, ,2007), interim targets that help to evaluate various scenarios (Lomas, 2010), and sectoral approaches that study consumption behaviours and production processes to achieve carbon emission

A. Cheshmehzangi and H. Chen, *China's Sustainability Transitions*,
https://doi.org/10.1007/978-981-16-2621-0_9

reductions (Tian et al., 2017). Moreover, the examples of hybrid modelling (Strachan & Kannan, 2008) suggest longer-term opportunities to consider carbon reduction, reducing demand, and other alternative modes of energy use, carbon capture and storage (CCS), etc. For instance, some industry-based programmes and technologies look at carbon reduction models (Fu et al., 2014), or other examples that explore costs, risks, and policy pathways (Scott et al., 2016), and investing in specific low carbon technologies (Gu & Wang, 2018) or experimentations (Cheshmehzangi et al., 2021; Cheshmehzangi, Xie, et al., 2018; Raven et al., 2019). However, there is often a policy clash (Bows & Anderson, 2007) that may lead to failures in experimentations. The combination of incentives and restrictions (Cheshmehzangi, 2014, 2016a) provides sustainability transition pathways, including low carbon targets to help have a better management plan and strategies that are not biased or one-sided. Nonetheless, in all cases, we see carbon reduction measures play a major part in sustainable urbanism, urban sustainability assessment, regional evaluations, and national targets. To highlight these further, we delve into specific examples of low carbon strategies.

In the following three sections, we look into three foremost examples of 'low carbon transport strategies', 'sustainable urban form and spatial planning', and 'renewable energies and smart grid'. These examples are identified as key areas of low carbon and climate-resilient urban research streams, which are highly relevant to China's low carbon transitions (Cheshmehzangi, Li, et al., 2018) and pathways towards the carbon-neutrality plan. By exploring these examples, we highlight several global examples. This chapter serves as the foundation of the next two chapters, focused on current progress (see Chap. 10) and future steps, lessons, and paradigms (see Chap. 11) for low carbon transitions in China.

9.2 Low Carbon Transport Strategies

9.2.1 Focused on: Transport and Connectivity: Strategy, Policy and Technology

In 2013, Suwon City in the Republic of Korea was the pilot city that held the inaugural Eco-mobility World Festival (EcoMobility, 2013; Suwon, 2013). Four years later, the event was held in the City of Kaohsiung, in Taiwan. The festival's preparation required resident surveys, expert input, rigorous planning, and the support of stakeholders and policy reinforcement. This festival led to the city of Suwon closing the streets of the Haenggung-dong area to fossil fuel-dependent automobiles (ICLEI, 2013; Moon et al., 2020). A similar approach was also developed and implemented in 2017 when Kaohsiung hosted the follow-up event. A pilot area was dedicated to the festival, enabling local people to achieve low carbon mobility and encourage new modes of urban travel. During planning in Suwon City, locals and business owners were worried that clients, visitors, and local residents themselves would not be able to access the area, thereby totally avoiding the areas. To address this issue, a range of workshops, consultation and training were created to educate, involve, and alleviate residents' concerns.

Moreover, to aid the transition, Suwon City held two car-free days in the months leading up to the festival and declared that four parking lots would be dedicated to the residents, accessed through electric shuttle vehicles (ICLEI, 2013; San, n.d.). This initiative provided the opportunity to revert the opinions of 70% of residents who initially opposed the event. The event also led to the physical transformation of the area; shop façades were repaired and revamped to make the main streets more inviting to pedestrians; streets were leveled and repaved, and former parking spaces and lots were designated as pocket parks, bicycle parking, and public spaces for residents to use and enjoy. The festival drew over 1 million visitors to Haenggung-dong. The same impact was also seen in the 2017 event. The event was further complimented by the weekend market and cultural activities. More than 20,000 visitors joined touring programs, and about 4300 residents adopted the Eco-mobile lifestyle. This included the use of a variety of bicycles (upright, tandem, recumbent), pedelecs (electric assisted bicycles), and velo-taxis (ICLEI, 2013). Also, closing major arterial streets to car traffic was met with resistance due to inconveniences to the commuter. Thus, it was advised to start with streets that cause fewer disruptions.

The alternative policy-based initiative used for low carbon development was the Eco-pass developed in the City of Milan (Haubold, 2015; Sgobbo & Basile, 2017; Sorrentino & Passerini, 2010). This is imposed as a selective charge to the most polluting vehicles. This initiative was executed in January 2008 to reduce traffic pollution. It was then replaced in 2012 with 'Area C policy', called a congestion charge (Percoco, 2013; Trivellato et al., 2019). For the Eco-pass, vehicles entering the area had to pay emissions charges proportional to their emission class (Mussone, 2017; Percoco, 2014). This is termed and is a type of cordon pricing, i.e., pricing or a fee related to the use of roadways defined by a cordon (Beria, 2016; Boggio & Beria, 2019). Another type is the congestion charge. For several years, the only example of such an initiative was in Singapore (Santos, 2005); but now similar pricing systems exist in London, Oslo, Stockholm, and Milan (Cheshmehzangi, Li, et al., 2018; Metz, 2018). Depending on the city's characteristics, the pricing system across different cities varies, and Milan tested the two various pricing schemes in short periods. Fundamentally, they looked at the low carbon effects between population charges and congestion charges. The Eco-pass targeted reducing PM10 concentration to keep in line with EU legislation (Cornes, 2016; Dameri, 2017; Jarl, 2009).

To begin this process, the government contacted a working group of academics and city officials to look at various cordon pricing options. The working group analysed previous examples of cordon pricing as done in Singapore, London, and Stockholm. This project led to two year trial of the Eco-pass initiative from 2008 to 2010 (Ruprecht & Invernizzi, 2009). Remarkably, the initiative reduced the PM10 concentration to below the European limit of 40 μg/m3. However, this reduction was reversed in 2011, and PM10 values reached 49 $\mu g/m^3$ (Goggi, 2011; Rahman et al., 2015). Two results were obtained from this initiative. Firstly, there was a shift from private cars to public buses (Croci, 2016). Secondly, the issue led to the increase of PM10 from the vehicles fleet renewal, which ended up increasing the number of vehicles entering the area (Martino & e Territorio, 2011; Pastorello et al., 2011; Beria, 2016). For the latter, the issues had an impact on increased emissions. The results of both initiatives were traffic reduction, public transport speed increase, air quality improvement, and revenue collection. Though initially focused on PM10 reduction, Eco-pass led to traffic reduction and composition (Cornes, 2016). The reduction was by 16% compared to 2007 levels, when no emission policy was in effect. Also, for traffic composition, low emission vehicles increased by 478%, and higher polluter vehicles declined by 48.1%. Additionally, Area C policy reduced traffic by 30.7% with respect to the last year of Eco-pass, amounting to 41,000 vehicles less per day. Also, road accidents reduced by 21.3% for eco-pass relative to 2007 levels and 23.8% for Area C with respect to 2011. In terms of air quality, Eco-pass reduced PM10 by 15% compared to the period without the Eco-pass. This was further reduced by 18% after the first year of Area C (Cheshmehzangi, Li, et al., 2018).

A more technologically-oriented initiative, which is needed for cities with a higher population, is methods of dealing with parking. A project that solved this problem was executed in San Francisco (Pierce & Shoup, 2013; Rodier & Shaheen, 2010). San Francisco's parking project provided real-time data on parking availability (Xu et al., 2013). This was done by using a demand response software mechanism to encourage parking in underutilised spaces. The overall aim was to reduce traffic congestion, increase circulation, and decrease air pollution from cruising cars. Thus, real-time data and demand-based pricing were used to ensure that there was one available parking spot per block at all times. In this project, multiple digital and physical methods were available for payment, and 8,200 sensors were used to provide the data to the mobile app or internet (Mathur et al., 2009; Zheng et al., 2015). Demand response pricing was particularly utilised to encourage under-utilised parking spaces. In fact, the pilot study showed that there was a 29% increase in revenue and 35% decrease in parking offences. This project was focused mainly on technology, automotive process, and real-time communication. In sum, the program helped increase parking availability, reduce parking search time and variability. It also reduced double parking, decreased long-term on-street parking, reduced congestion, improved speed and reliability of public transit, improved air quality, and reduced greenhouse gas emission. The longer-term effect was the improved economic vitality of pilot areas, which helped augment San Francisco Municipal Transport's financial sustainability and Improve transit, taxi, pedestrian, bicyclist, and driver safety.

Here, we summarise 16 key lessons learned for the context of China. First, we refer to the importance of consultation with residents—i.e., seeking comments, suggestions, and feedback in the process. This is seen as an important participatory approach to bringing back people in decision-making procedures (Cheshmehzangi & Dawodu, 2019a). Second is the availability of legal mechanisms, which do not always have to be used for low carbon development projects. These may sometimes be appealing to the residents, and they may provide more efficient approaches to optimise the organisation of low carbon projects. Third is the gradual introduction of low carbon initiatives to the general public, something that appears more effective in multiple steps or stages of development (2020c; Cheshmehzangi, 2014). Fourth, we stress the role of mobilising local champions in support of initiatives. The role of multiple stakeholders and their position in quality improvement (Kirchner et al., 2012) is recognised as a successful model of achieving holistic decisions and strategies that enable better transitions. Fifth, we highlight the importance of low carbon policies/initiatives, which should be pilot-tested before full implementation (Cheshmehzangi, Xie, et al.,2018). Sixth, a combination of policies that complement each other via strategies to low carbon development could be seen as a holistic model of low carbon development. The idea is to capitalise people's shift to public vehicles or move away from unsustainable modes or behaviours. Seventh, we refer to a regular review of policy initiatives, which is required due to the ever-changing response by people. In doing so, we are able to upgrade the standards and benchmarks continuously, too. Eighth, we highlight the fact that research on the implication of low carbon policies needs to be carried out before being put forward for implementation. This stage is seen as vital to developing urban sustainability indicators and low carbon transitions. Ninth, we note that communication and transparency are key in maintaining good relationships between different departments, stakeholders, and actors. The multi-stakeholder constellation mechanism enables better opportunities for healthier transitions and sustainability pathways (Deng & Cheshmehzangi, 2018). Tenth, we highlight approaches to support alternative infrastructure that take into consideration demographic location and providing/ improving alternative forms of low carbon transport. The 11th lesson is to explore low carbon development, which can be optimised through innovative technologies and their applications (Albino et al., 2014; Köhler et al., 2013; Lyu et al., 2020; Mercure et al., 2019; Polzin, 2017; Shi & Lai, 2013). The 12th lesson is the use of real-time data, which should be used for low carbon projects such as energy, transport, waste, etc. In the same vein, feedback data on more strategic and policy-driven low carbon approaches is also paramount (Bakker et al., 2017). The 13th lesson is to consider low carbon projects for environmental benefits and address social and economic issues of the region. In doing so, we could shift towards more holistic examples of low carbon initiatives, programmes, policies, etc. The 14th lesson is the consideration of feasibility studies that incorporates stakeholder surveys (Cheshmehzangi, 2020c). This consideration is essential to understand the dynamics of a new location, specifically by exploring the opportunities that may be only context-specific. Our last lesson for China is to consider reinvesting a portion of the financial benefits of low carbon initiatives, which should be reinvested back to improve living conditions. The financing aspect is a critical aspect (Zhan & de Jong, 2018), which should be resolved through sustainable mechanisms of revenue generation and benefiting the initiatives in a long term.

9.3 Sustainable Urban Form and Spatial Planning

9.3.1 Focused on: Green Infrastructure Strategies and Policies

Green infrastructure (GI) is a low carbon approach used to address urban heat island effect issues, provide a cool environment, and mitigate the pollution. It also helps to illustrate visually the efforts being made to keep a city sustainable (Butters et al., 2020; Cheshmehzangi & Butters, 2017; Gaffin et al., 2012; Koc et al., 2017; van der Jagt et al., 2019). Such an approach is the city of Baguio in the Philippines, with benefits reaped from participatory approach to planning, not only environmental but also economic and social (Estoque & Murayama, 2013). The region of Baguio has maintained the uniqueness of flora and fauna, which can only be found in the city and contributes to a thriving tourism economy (Baoanan et al., 2020). Due to population growth, urban sprawl (Gonzales, 2016), and deforestations (Estoque & Murayama, 2011, 2012), these biodiverse benefits were initially put under jeopardy. However, as early as the 1960s to the 1970s, the city realised the need to build citizens' consciousness to biodiversity preservation and enhancement of cities natural beauty through flower gardens and tree planting. This initiative motivated a group of professionals to form the Baguio City Parks Foundation (BCFF) to assist the government in managing and maintaining parks and playgrounds (Estoque & Murayama, 2013). This partnership's success led to the adopt the park project in 1989 (Chua & Scura, 1991) to further attract civic-minded organisations to safeguard, maintain, and develop parks and gardens in the city. This approach also spurred local villages to designate public land areas for community parks and villages, which were given approval by the local council to protect, maintain, and enhance their water resources and biodiverse environment. By the government soliciting help from the citizens and civic organisations and involving them in the process, this project became the catalyst to the success of many of the green initiatives, e.g., the 'Eco Walk' (Dacawi, 1999; Mercado, 1998). The program targeted experiential learning in order to raise consciousness. The activities included hikes to the watershed, observation tours of flora and fauna, lectures on the effect of squatting, waterlogging, resource depletion, and climate change, and quizzes on the environment and tree planting. This led to the planting of 25,000 trees. There were also strong volunteers within the private sector, local government, NGOs, churches, and academia leading to the project's success. In sum, it is evident that enabling stakeholder involvement and their recognition and ownership in the project improved this policy-led approach to success.

In a similar yet different vein, the City of Adelaide, Australia, has also committed to a greener livable city (Lam & Mullen, 2012; Li et al., 2017; Nouri et al., 2019, 2020). However, they took the vertical route instead of the horizontal route (Bustami et al., 2019; Razzaghmanesh et al., 2016). Green walls were developed over the Council's Customer Centre building on Pirie Street in Adelaide. This is in addition to cascading wall of 2,330 plants over the redesigned plaza, which provides integrated green bike racks, landscaping, and improved access council plaza area. The vertical project is the city's attempt to infuse and increase greenery into built-up areas, especially those commercial areas and business districts. This approach's main indication was to show business owners, vendors, regular commuters, and shoppers how GI can be integrated into building façade (Perrotti & Iuorio, 2019). The general idea is to add to the city's horizontal greenery by the provision of more trees and plants for every building, thereby further cooling the environment, improving air quality, and adding vibrancy to public spaces (Sharifi et al., 2016). It also addresses places that are bare and blanks and not performing environmentally, essentially making the entire environment low carbon.

Another interesting approach to low carbon development from a policy perceptive but utilising 'greening' or GI (in a different manner) is the 'one day per week vegetarian school lunch policy' by Taichung City. The project focused on local schools' switching dietary habits of students to address land availability, reduce carbon emissions, and create a local sustainable economic sector (Huang et al., 1999). The logic is that food without meat can save up to 0.78 kg of CO_2 emissions (Wu, 2014). It is more energy-efficient and financially sustainable to produce vegetables than to produce or import meat, especially with limited farmland, water, and fossil fuel (Cheshmehzangi & Dawodu, 2019b). The schools were targeted not only to raise awareness at early age low carbon behaviors but because they are one of the largest public bodies in cities. The government's approach was to improve sustainable procurement and low carbon development by tailoring the option to healthier and environmentally beneficial options. In order to replicate such a process, the first principle was to start to slow. This meant making the initiative voluntary before it is made mandatory (Cheshmehzangi, Li, et al., 2018). Besides, training, consultation, and proper awareness programs were included for people to understand the benefits of such initiatives. These approaches were included since the project was developed in a very top-down manner, with minimal public say on the policy (Cheshmehzangi & Dawodu, 2019b). The final step was patience, which means that changing dietary habits is a gradual operation that requires time. It also requires a gradual reduction in old habits (such as processed and meat-orient food), to be replaced by a new healthier lifestyle. It also helps that the younger generation of the future generations was targeted and are more malleable to change.

Another international example is the City of Tokyo, Japan. In this project, a patchwork of various urban spaces is provided, allowing for spatial diversity and detailed neighbourhood/district planning. A similar approach is also replicated in other major cities, like London. The Tokyo project indicates a distribution of diverse types of urban spaces (Imai, 2017), including key strategies and detailed planning such as (1) prevention of urban sprawl and protection of natural resources and farmlands, (2) efficient transit system with effective land-use strategies, (3) intensification of urban centers and redevelopment of brownfields, (4) conservation of historic areas, (5) improvement of informal (popular) settlements, and (6) maintenance and improvement of the suburbs. In this project, we see the importance of spatial planning and urban form, similar to those seen in urban regeneration projects in Australia's greyfield suburbs (Newton, 2010), industrial zones (Tiesdell, 2008), and small to medium-sized cities in the UK and Canada (Millward, 2006). The nexus between urban form and sustainability in Tokyo (Okata & Murayama, 2011) shows the importance of space variety, their distributions, accessibility, and improvements. These projects could be seen as examples of low carbon development, utilising spaces, and cities' public realms.

Here, we summarise seven key lessons Learned for the context of China. First, if given the opportunity, we recommend relevant stakeholders (i.e., in a mixed group) serve as efficient decision-makers and planners. We could achieve an integrated approach (both Top-down and Bottom-up) to low carbon development, which ultimately optimises the planning approach and subsequent results. Second, we believe government support through delegation of task, incentives, and policies (Deng et al., 2020; Huxham et al., 2019; Jagger et al., 2013; Roy et al., 2013; Zhang et al., 2014, 2019) is essential in achieving a sustainable low carbon transition. The approach to enhance financial systems (Campiglio et al., 2017) helps expand government revenues and diverse effects on low carbon opportunities. Third is innovation in practice and policy development, which should be looked at from the perspective of power relations (Tyfield et al., 2015) and systematic development of regional sustainable development (Yin et al., 2019), and socio-technical transitions (Geels et al., 2018). Fourth is targeting future generations leading to long-term solutions (Cheshmehzangi & Dawodu, 2019b), which is the backbone of longer-term investment and development, such as in low carbon development. Fifth is the provision of training, consultation, and raising awareness, which are all needed for low carbon ideals to be truly welcomed and imbibed. Sixth, we refer to low carbon development, which should be supported by other economic and social benefits such as tourism. Fundamentally, low carbon development should be sustainable, and its realisation in other sectors is also highly important (He and Wang, 2021). Examples of market integration (Gössling, 2003), management perspectives (Menendez-Carbo et al., 2020), and technology innovation and applications (Ali & Frew, 2014) are successful methods of achieving sustainable models of low carbon development. Lastly, we note the competition is healthy and, if properly incentivised, can optimise the low carbon development performance. A health competition could help also develop better efficiency and investment mechanisms (Chen et al., 2017), which would lead to better low carbon economy transitions (Hu & Liu, 2010) and quality development.

9.4 Renewable Energies and Smart Grid

9.4.1 Focused on: Renewable Energies: Technology, Policies and Targets

In terms of renewable energy (RE) and Energy Efficiency (EE), several technologies can achieve the aims of low carbon emissions (Yuan et al., 2011; Wang & Chang, 2014; Zhou et al., 2014, 2017; Jia et al., 2018). Currently, China has invested in myriad or combination of them, divided into supply and demand approaches (Baeumler et al., 2012). The supply approach involves generating energy from solar, wind, geothermal, hydro, or biomass energy sources. Another approach is the demand reduction strategy, which involves minimising energy used by improving technology efficiency (Qu & Liu, 2017). This approach is currently utilised in Nagpur, India (Sukhwani et al., 2020), and Sydney, Australia (Trencher & Van der Heijden, 2019). Both cities' strategies have involved replacing existing lighting with more energy-efficient lighting systems (e.g., LED lights), powering lighting through standalone PV. This approach helps to reduce energy consumption from the grid. Moreover, the grid-interactive PV-powered lighting involves a system that contributes electricity to the grid by day when lights are off and draws from it at night. This system differs from the standalone systems, ensuring that all power produced is used, thereby avoiding lighting failure (Koko et al., 2018; Sharma & Chandel, 2013).

Another strategy is grid-interactive PV-powered lighting, and this is a smart grid technology that uses real-time data and communication technology (Nelson et al., 2017; Peng et al., 2013). This means that the lights are switched on when natural daylight falls below a minimum threshold (Dhar et al., 2017). Another method of requiring smart grid technology is the procurement of RE lighting. This means purchasing the needed electricity and feeding it into the grid from a clean energy source (Matthäus, 2020; Miller, 2020). These objectives of development included (1) establishing a renewable resource center to raise awareness, (2) identifying relevant stakeholder groups, and (3) providing training programs and awareness-raising events to various stakeholders. The last point brings forward a critical aspect that China can address in terms of RE and EE. This aspect requires the setting of local targets to establish the direction of current and future actions and support national targets' achievement. The approach is to set more comprehensive and ambitious targets for RE and EE. In order to establish such targets, it is essential to have the knowledge and understanding of the context and local energy situation (Cai et al., 2009; Ma et al., 2009; Zhang et al., 2017). This means knowing what RE resources are available, the human resources available to exploit the RE resource, the region's technology level, etc. An energy mapping or potential analysis would be needed to achieve these, along with any other survey deemed necessary. For the lighting case of Sydney, the lighting target was to reduce 70% of GHG emissions by 2030 based on 2006 levels; to remove reliance on coal-based electricity by 2030; 70% of the local government's electricity needs should come from tri-generation (combined production of cooling, heat, and power), and 30% from renewable electricity. The key improvements were achieving a 20% reduction of GHG emissions in 2012 compared to the figures of local government operations in 2006/2007. For Nagpur, after developing the renewable resources center, the RE target stipulated a 3% reduction in conventional energy consumption across the city. An additional target was the 20% reduction in conventional energy consumption in municipal operations and facilities. This made the city of Nagpur become India's first model solar city (Tiwari & Kalamkar, 2016; Werulkar & Kulkarni, 2015) and made them eligible for national funds.

A key strategy used in both Sydney and Nagpur was introducing ambitious targets for their municipal operations. Such targets cover buildings, technologies, utilities, vehicles, equipment, etc. Since this is usually under the local government administration, the use of RE and EE strategies would be easy and more rapidly executed. The next important aspect is that RE and EE strategies can be made for communities or the entire city by integrating the plans into the city masterplan strategy document. In addition, targets can be executed according to sectors, e.g., building, transport, and lighting. A city that has taken these initiatives to the heart is the City of Malmö, Sweden. They have their ambitious strategy outlined in two documents, the Environmental Program and the Energy Strategy (Austin, 2013; Bibri & Krogstie, 2020; Stripple & Bulkeley, 2019). The strategy was for the city as a whole and municipal operations, and aim that by 2020, the city of Malmö will be climate neutral, and by 2030 all municipal operations will run on 100% renewable energy (Bulkeley & Stripple, 2020; Lenhart et al., 2014; Parks, 2020). The most considerable contribution is to come from locally generated RE, and these objectives by far superseded the aim of the European Union plan Sweden (49% by 2020).

Here, we summarise 12 key lessons Learned for the context of China. First, based on global examples' success story, we believe ambitious targets should be set by local regions that supersede national targets. The adaptation and localisation of policies are essential in achieving low carbon targets (Deng et al., 2020). In the case of China, this would mean superseding the energy targets of the five-year-plans. Second, low carbon targets should link supply and demand strategies to RE development. Third, it is evidenced that RE solutions should be tailored to the local resources and conditions (e.g., human power, local climate, technologies, and skill level). Fourth, it is necessary to have legally or politically binding documents, which should be infused with city or community master plans. Fifth, the masterplan development should have action plans, strategies, and targets for RE and EE execution (Cheshmehzangi et al., 2021; Cheung et al., 2019; Giaccone et al., 2017; Poggi et al., 2017). Sixth, we refer to associated costs and lifecycle assessment of the RE and EE, which are necessary to be considered to optimise maintenance and lifetime low carbon technologies. In doing so, we are able to make RE and EE initiatives more affordable and accessible to a broader range of users. Seventh, we highlight the importance of raising awareness for policymakers, the public, and other relevant stakeholders are pivotal to RE and EE development and target setting. Eighth, RE targets should encompass the whole community and be based upon a participatory approach (Adams et al., 2011; Schmid & Knopf, 2012). Ninth, we suggest trials and pilot studies of RE technologies and policies are needed in order to optimise large-scale implementations for scaling up projects (Beermann & Tews, 2015; Mabrouk et al., 2015). Tenth, we refer to the local policies that need to support RE strategies and feed into national policies and targets. To achieve this successfully, political leader(s) should be chosen to push the RE agenda forward. The 11th lesson is a suggestion for local governments to start implementing RE and EE strategies with municipal operations and leading an example. Lastly, we highly recommend local governments to set the targets together with energy companies as key actors and drivers for transition.

9.5 Current Status of China's Low Carbon Development: Overview of the Recent Progress

As we enter the new development phase in China, it is important to highlight its recent progress. The "13th FYP" period is regarded as the phase of complete success in building a well-to-do society. This idea came to recognition much earlier (Chen, 2004), suggesting the importance of this Chinese theory in the practice of urbanisation and development (Cheshmehzangi, 2016b; Sun et al., 2020). According to the clear deployment of the party's 19th National Congress, it is necessary to comprehensively inspect the implementation of the main goals and 25 key indicators of the country's "13th FYP" (Xi, 2017). The other actions are proposed to evaluate the progress made in implementing the plan objectively, summarise the experience and practices for refining and promoting the implementation of the plan, and analyse in-depth the problems and causes arising from the implementation (Cheshmehzangi, Li, et al., 2018). The intention was to combine the changes in the development environment at home and abroad, propose countermeasures and suggestions for improving the implementation of the plan, and further strengthen the national development plan's strategic orientation (ibid). In doing so, China can continue promoting the smooth implementation of various tasks, ensuring that the well-to-do society is fully completed on time. This is believed to lay a more solid foundation for starting a new journey to build a socialist modernisation country (ibid). Therefore, in the following chapter, we briefly review the 'Low Carbon Development Policy' outline during the 13th five-year plan (FYP) period.

As we have highlighted in previous chapters, green and low carbon development is an important part of the 13th FYP, which involves various goals such as economic development, optimisation of industries and energy structures, and improvement of the ecological environment's quality, and improvement of people's quality of life. In order to ensure the smooth implementation of the 13th FYP, key sectors such as energy, construction, and other fields respectively issued the "13th FYP for Energy Development", "13th FYP for Development of the Construction Industry," and "13th FYP for Controlling Greenhouse Gas Emissions". In addition, other series of supporting special plans and policies formulated clear development goals and task measures (ibid). The assessment of the effectiveness of these special plans and policies can not only provide recommendations for the overall assessment of the 13th FYP but also can systematically study the contribution of these special plans and policies to China's contribution to green and low carbon development and climate change. In doing so, China could sum up experiences, find shortcomings, analyse causes, propose countermeasures, and pay attention to digging deep-seated contradictions and hidden dangers of the recent past. This helped the country discover new problems and contemporary issues and make clear plans to implement the second half of the critical tasks and requirements (ibid). Therefore, we need to evaluate China's progress in further detail and see what could be suitable for future low carbon development strategies, directions, policies, and practices. Nonetheless, the low carbon futures are structures in overarching plans of industrial restructuring, decarbonisation plan, and high-quality urbanisation. These plans would require effective low carbon initiatives, integrated and multi-objective strategies, and low carbon urban development. We will cover these factors in the remaining chapters of the book.

9.6 Summary

The main measures for green and low carbon development include (1) adjustment of industrial structure, (2) optimisation of energy structure, (3) energy conservation and improvement of energy efficiency, (4) development of low carbon transportation and construction, and (5) promotion of low carbon lifestyles. Among them, optimising energy structure and vigorously developing renewable energy is considered the most important way to reduce CO_2 emissions (ibid). Therefore, The Chinese National Government introduced several policies in the first two years of the 13th FYP to reduce fossil energy consumption and greatly encourage the development of the renewable energy industry. As per records, these policies have achieved very significant results (ibid). Measures to increase energy efficiency and the development of building energy efficiency and smart grids also contribute significantly to reducing carbon emissions. Therefore, in the next chapter we focus on four key areas of (1) energy consumption, (2) renewable energy, (3) energy efficiency in buildings and construction, and (5) smart grids. We evaluate these to better understand China's progress and key low carbon indicators. In doing so, we are able to analyse problems and challenges and put forward some necessary recommendations for the 2030 targets and the 2060 plan.

In the next two chapters, we explore some of the main schemes developed or developing to promote and/or achieve low carbon transitions. In doing so, we explicitly focus on recent low carbon progress in China and potential low carbon futures, such as lessons and paradigms that are specifically relevant to China's context. Some of these lessons and paradigms could also be relevant to rapidly-developing countries or the contexts where low carbon and climate-resilient agendas are considered the top priority.

Our conclusion of this chapter is mainly based on considering low carbon strategies and learning from them. By learning from mistakes and challenges and reflecting on some of the key directions, we could highlight the role of mainstreaming low carbon strategies in future urban development and sustainability transitions. As we delve into more details of China's recent progress (Chap. 10), we highlight some of the strengths, weaknesses, opportunities, and threats/challenges. We look into two sides of the coin, and in particular, explore what could be learned from other examples, from China's challenges, and potential contextualised paradigms. The importance of policies and practices is highlighted in the book's remaining chapters, suggesting pathways to achieving the 2030 carbon targets and the 2060 carbon neutrality goal.

Acknowledgments Ali Cheshmehzangi acknowledges the Asian Development Bank (ADB) and the National Development Reform Commission (NDRC) of China for the opportunity of a research project that he led during 2017 and 2018. In particular, we acknowledge colleagues and friends from ADB for their support and the opportunity to review China's 13th Five-Year-Plan, which then identified the critical areas of low carbon development for the 14th Five-Year-Plan. Besides, we thank our team members at 'Chinese Society for Urban Studies (CSUS)' and 'National Center for Climate Change Strategy and International Cooperation (NCSC)', both based in Beijing, China. In particular, we acknowledge the support of Dr. Hailong LI (leading the CSUS Team) and Dr. Xiu YANG (leading the NCSC Team). With their support and valuable collaboration, we managed to put together a very detailed review and a comprehensive technical report.

References

Adams, M., Wheeler, D., & Woolston, G. (2011). A participatory approach to sustainable energy strategy development in a carbon-intensive jurisdiction: The case of Nova Scotia. *Energy Policy, 39*(5), 2550–2559.

Albino, V., Ardito, L., Dangelico, R. M., & Petruzzelli, A. M. (2014). Understanding the development trends of low-carbon energy technologies: A patent analysis. *Applied Energy, 135*, 836–854.

Ali, A., & Frew, A. J. (2014). Technology innovation and applications in sustainable destination development. *Information Technology & Tourism, 14*(4), 265–290.

Austin, G. (2013). Case study and sustainability assessment of Bo01, Malmö, Sweden. *Journal of Green Building, 8*(3), 34–50.

Bakker, S., Dematera Contreras, K., Kappiantari, M., Tuan, N. A., Guillen, M. D., Gunthawong, G., Zuidgeest, M., Liefferink, D., & Van Maarseveen, M. (2017). Low-carbon transport policy in four ASEAN countries: Developments in Indonesia, the Philippines, Thailand and Vietnam. *Sustainability, 9*(7), 1217

Baeumler, A., Ijjasz-Vasquez, E., & Mehndiratta, S. (Eds.). (2012). *Sustainable low-carbon city development in China.* World Bank Publications.

Bagheri, M., Delbari, S. H., Pakzadmanesh, M., & Kennedy, C. A. (2019). City-integrated renewable energy design for low-carbon and climate-resilient communities. *Applied Energy, 239*, 1212–1225.

Baoanan, Z. G., Abansi, C. L., & Abalos, C. F. S. (2020). A review of biodiversity-related reports in the Cordillera Highlands, Northern Luzon, Philippines. *Journal of Nature Studies, 19*(2), 84–103.

Beermann, J., & Tews, K. (2015). Preserving decentralised laboratories for experimentation under adverse framework conditions: Why local initiatives as a driving force for Germany's renewable energy expansion must reinvent themselves.

Beria, P. (2016). Effectiveness and monetary impact of Milan's road charge, one year after implementation. *International Journal of Sustainable Transportation, 10*(7), 657–669.

Bibri, S. E., & Krogstie, J. (2020). Smart eco-city strategies and solutions for sustainability: The cases of Royal Seaport, Stockholm, and Western Harbor, Malmö, Sweden. *Urban Science, 4*(1), 11.

Boggio, M., & Beria, P. (2019). The role of transport supply in the acceptability of pollution charge extension. The case of Milan. *Transportation Research Part A: Policy and Practice, 129*, 92–106.

Bows, A., & Anderson, K. L. (2007). Policy clash: Can projected aviation growth be reconciled with the UK Government's 60% carbon-reduction target? *Transport Policy, 14*(2), 103–110.

Bulkeley, H., & Stripple, J. (2020). Climate smart city: New cultural political economies in the making in Malmö, Sweden. *New Political Economy*, 1–14.

Bustami, R. A., Brien, C., Ward, J., Beecham, S., & Rawlings, R. (2019). A statistically rigorous approach to experimental design of vertical living walls for green buildings. *Urban Science, 3*(3), 71.

Butters, C., Cheshmehzangi, A., & Sassi, P. (2020). Cities, energy and climate: Seven reasons to question the dense high-rise city. *Journal of Green Building, 15*(3), 197–214.

Cai, W. G., Wu, Y., Zhong, Y., & Ren, H. (2009). China building energy consumption: Situation, challenges and corresponding measures. *Energy Policy, 37*(6), 2054–2059.

Campiglio, E., Godin, A., Kemp-Benedict, E., & Matikainen, S. (2017). The tightening links between financial systems and the low-carbon transition. In *Economic policies since the global financial crisis* (pp. 313–356). Palgrave Macmillan, Cham.

Chen, Y. (2004). The study on establishing assessment indicator system about construction of the well-to-do society. *Sociological Research, 1*. (in Chinese).

Chen, X., Luo, Z., & Wang, X. (2017). Impact of efficiency, investment, and competition on low carbon manufacturing. *Journal of Cleaner Production, 143*, 388–400.

Cheshmehzangi, A. (2014). The urban and urbanism: China's new urbanization and approaches towards comprehensive development. *The International Journal of Interdisciplinary Environmental Studies, 8*(3–4), 1–12.

Cheshmehzangi, A. (2016). Feasibility study of Songao's low carbon town planning, China. *Energy Procedia, 88*, 313–320.

Cheshmehzangi, A. (2016). China's New-Type Urbanisation Plan (NUP) and the foreseeing challenges for decarbonization of cities: A review. *Energy Procedia, 104*, 146–152.

Cheshmehzangi, A. (2020). Towards urban energy management: A brief review of China's recent transition in urban energy demand. *Low Carbon Economy, 11*(1), 1–24.

Cheshmehzangi, A. (2020). Evaluating the Nexus between housing and energy sectors: The comparison of urban, Peri-urban and rural housing areas in Zhuhai, China. *Energy and Power Engineering, 12*(06), 314.

Cheshmehzangi, A. (2020c). Low carbon transition at the township level: Feasibility study of environmental pollutants and sustainable energy planning. *International Journal of Sustainable Energy*, 1–27. https://doi.org/10.1080/14786451.2020.1860042.

Cheshmehzangi, A., & Butters, C. (Eds.). (2017). *Designing cooler cities: Energy, cooling and urban form: The Asian perspective*. Palgrave Macmillan.

Cheshmehzangi, A., & Dawodu, A. (2019a). The review of sustainable development goals (SDGs): People, perspective and planning. In *Sustainable urban development in the age of climate change* (pp. 133–156). Palgrave Macmillan, Singapore.

Cheshmehzangi, A., & Dawodu, A. (2019b). Case study reviews: People, perspective and planning. In *Sustainable urban development in the age of climate change* (pp. 69–131). Palgrave Macmillan.

Cheshmehzangi, A., Dawodu, A., & Sharifi, A. (2021). *Sustainable urbanism in China*. Routledge.

Cheshmehzangi, A., Xie, L., & Tan-Mullins, M. (2018). The role of international actors in low-carbon transitions of Shenzhen's International Low Carbon City in China. *Cities, 74*, 64–74.

Cheshmehzangi, A., Li, H., & Yang, X. (2018, October). Low carbon and climate resilient urban development in China. ADB Report, developed for the National Development Reform Committee (NDRC), Beijing, China.

Cheung, G., Davies, P. J., & Trück, S. (2019). Transforming urban energy systems: The role of local governments' regional energy master plan. *Journal of Cleaner Production, 220*, 655–667.

Cheng, Y., & Yao, X. (2021). Carbon intensity reduction assessment of renewable energy technology innovation in China: A panel data model with cross-section dependence and slope heterogeneity. *Renewable and Sustainable Energy Reviews, 135*, 110157.

Chua, T. E., & Scura, L. F. (1991). *Managing ASEAN's Coastal Resources for Sustainable Development: Roles of Policymakers, Scientists, Donors, Media and Communities: Proceedings of the ASEAN/US Policy Conference on Managing ASEAN's Coastal Resources for Sustainable Development, Manila and Baguio, Philippines, 4–7 March 1990* (Vol. 6). WorldFish.

Cornes, C. (2016). *Effectiveness of urban mobility management plans to improve air quality* (Master's thesis, Universitat Politècnica de Catalunya).

Croci, E. (2016). Urban road pricing: A comparative study on the experiences of London, Stockholm and Milan. *Transportation Research Procedia, 14*, 253–262.

Dacawi, R. S. (1999). Media and good governance: a case study. Available from https://dr.ntu.edu.sg/bitstream/10220/2397/1/AMIC_1999_NOV10-12_12.pdf

Dameri, R. P. (2017). ICT intensity in smart mobility initiatives. In *Smart city implementation* (pp. 85–108). Springer.

Dawodu, A., & Cheshmehzangi, A. (2017). Impact of floor area ratio (FAR) on energy consumption at meso scale in China: Case study of Ningbo. *Energy Procedia, 105*, 3449–3455.

Deng, W., & Cheshmehzangi, A. (2018). *Eco-development in China: Cities, communities and buildings*. Palgrave Macmillan.

Deng, W., Cheshmehzangi, A., Ma, Y., & Peng, Z. (2020). Promoting sustainability through governance of eco-city indicators: A multi-spatial perspective. *International Journal of Low-carbon Technologies*. 16(1), 61–72.

Dhar, S., Patnaik, R. K., & Dash, P. K. (2017). Fault detection and location of photovoltaic based DC microgrid using differential protection strategy. *IEEE Transactions on Smart Grid, 9*(5), 4303–4312.

Ding, L. L., Lei, L., Zhao, X., & Cantemir Calin, A. (2020). Modelling energy and carbon emission performance: A constrained performance index measure, *Energy, 197*, 117274.

Druckman, A., Bradley, P., Papathanasopoulou, E., & Jackson, T. (2008). Measuring progress towards carbon reduction in the UK. *Ecological Economics, 66*(4), 594–604.

EcoMobility. (2013). Suwon Congress Report, ICLEI—Local Governments for Sustainability, Report.

Estoque, R. C., & Murayama, Y. (2011). Spatio-temporal urban land use/cover change analysis in a hill station: The case of Baguio City, Philippines. *Procedia-Social and Behavioral Sciences, 21*, 326–335.

Estoque, R. C., & Murayama, Y. (2012). Examining the potential impact of land use/cover changes on the ecosystem services of Baguio city, the Philippines: A scenario-based analysis. *Applied Geography, 35*(1–2), 316–326.

Estoque, R. C., & Murayama, Y. (2013). City profile: Baguio. *Cities, 30*, 240–251.

Fu, J. X., Tang, G. H., Zhao, R. J., & Hwang, W. S. (2014). Carbon reduction programs and key technologies in global steel industry. *International Journal of Iron and Steel Research, 21*(3), 275–281.

Gaffin, S. R., Rosenzweig, C., & Kong, A. Y. (2012). Adapting to climate change through urban green infrastructure. *Nature Climate Change, 2*(10), 704–704.

Geels, F. W., Schwanen, T., Sorrell, S., Jenkins, K., & Sovacool, B. K. (2018). Reducing energy demand through low carbon innovation: A sociotechnical transitions perspective and thirteen research debates. *Energy Research & Social Science, 40*, 23–35.

Giaccone, A., Lascari, G., Peri, G., & Rizzo, G. (2017). An ex post criticism, based on stakeholders' preferences, of a residential sector's energy master plan: the case study of the Sicilian region. *Energy Efficiency, 10*(1), 129–149.

Gil, L., & Bernado, J. (2020). An approach to energy and climate issues aiming at carbon neutrality. *Renewable Energy Focus, 33*, 37–42.

Goggi, G. (2011). Ecopass: quali risultati. *Focus, 47*(2).

Gonzales, L. B. F. (2016). Urban sprawl: Extent and environmental impact in Baguio City, Philippines. *Spatium*, 7–14.

Gössling, S. (2003). Market integration and ecosystem degradation: Is sustainable tourism development in rural communities a contradiction in terms? *Environment, Development and Sustainability, 5*(3), 383–400.

Gu, G., & Wang, Z. (2018). Research on global carbon abatement driven by R&D investment in the context of INDCs. *Energy, 148*, 662–675.

Haubold, H. (2015). Congestion charges and cycling. Available from https://repository.difu.de/jspui/bitstream/difu/242709/1/DS1679.pdf.

He, J., & Wang, D. (2021, January). On low-carbon and its realization in tourism. In *IOP conference series: Earth and environmental Science* (Vol. 632, No. 5, p. 052102). IOP Publishing.

Hu, Z. W., & Liu, Y. (2010). Provincial competition study from the perspective of low-carbon economy. *China Industrial Economics, 4*. (in Chinese).

Huang, Y. C., Lin, W. J., Cheng, C. H., & Su, K. H. (1999). Nutrient intakes and iron status of healthy young vegetarians and nonvegetarians. *Nutrition Research, 19*(5), 663–674.

Huxham, M., Anwar, M., & Nelson, D. (2019). Understanding the impact of a low carbon transition on South Africa. *Climate Policy Initiative (CPI).* https://climatepolicyinitiative.org/wpcontent/uploads/2019/03/CPI-Energy-Finance-Understanding-the-impact-of-a-low-carbon-transition-South-Africa-March-2019.pdf.

ICLEI (Local Government for Sustainability). (2013). *EcoMobility world festival 2013*, 12 pages document. Available from https://www.sustainabledevelopment.un.org.

Imai, H. (2017). *Tokyo roji: the diversity and versatility of alleys in a city in transition.* New York and London: Routledge.

Ismailos, C., & Touchie, M. F. (2017). Achieving a low carbon housing stock: An analysis of low-rise residential carbon reduction measures for new construction in Ontario. *Building and Environment, 126*, 176–183.

Jagger, N., Foxon, T., & Gouldson, A. (2013). Skills constraints and the low carbon transition. *Climate Policy, 13*(1), 43–57.

Jarl, V. (2009). Congestion pricing in urban areas: Theory and case studies. Available from https://lup.lub.lu.se/luur/download?func=downloadFile&recordOId=8920835&fileOId=8920836.

Jia, P., Li, K., & Shao, S. (2018). Choice of technological change for China's low-carbon development: Evidence from three urban agglomerations. *Journal of Environmental Management, 206*, 1308–1319.

Kairies-Alvarado, D., Muñoz-Sanguinetti, C., Martínez-Rocamora, A. (2021). Contribution of energy efficiency standards to life-cycle carbon footprint reduction in public buildings in Chile, *Energy and Buildings, 236*, 110797.

Kellett, J. (2007). Community-based energy policy: A practical approach to carbon reduction. *Journal of Environmental Planning and Management, 50*(3), 381–396.

Kirchner, J. E., Parker, L. E., Bonner, L. M., Fickel, J. J., Yano, E. M., & Ritchie, M. J. (2012). Roles of managers, frontline staff and local champions, in implementing quality improvement: Stakeholders' perspectives. *Journal of Evaluation in Clinical Practice, 18*(1), 63–69.

Koc, C. B., Osmond, P., & Peters, A. (2017). Towards a comprehensive green infrastructure typology: A systematic review of approaches, methods and typologies. *Urban Ecosystems, 20*(1), 15–35.

Köhler, J., Schade, W., Leduc, G., Wiesenthal, T., Schade, B., & Espinoza, L. T. (2013). Leaving fossil fuels behind? An innovation system analysis of low carbon cars. *Journal of Cleaner Production, 48*, 176–186.

Koko, S. P., Kusakana, K., & Vermaak, H. J. (2018). Optimal power dispatch of a grid-interactive micro-hydrokinetic-pumped hydro storage system. *Journal of Energy Storage, 17*, 63–72.

Laes, E., Mayeres, I., Renders, N., Valkering, P., & Verbeke, S. (2018). How do policies help to increase the uptake of carbon reduction measures in the EU residential sector? Evidence from recent studies. *Renewable and Sustainable Energy Reviews, 94*, 234–250.

Lam, A. H., & Mullen, B. M. (2012). comparative analysis of best practices of sustainable communities: Adelaide, Australia case study. *Cityscape*, 235–242.

Lenhart, J., Bouteligier, S., Mol, A. P., & Kern, K. (2014). Cities as learning organisations in climate policy: The case of Malmö. *International Journal of Urban Sustainable Development, 6*(1), 89–106.

Li, F., Sutton, P. C., Anderson, S. J., & Nouri, H. (2017). Planning green space in Adelaide city: Enlightenment from green space system planning of Fuzhou city (2015–2020). *Australian Planner, 54*(2), 126–133.

Li, X., & Yao, X. (2020). Can energy supply-side and demand-side policies for energy saving and emission reduction be synergistic? A simulated study on China's coal capacity cut and carbon tax, *Energy Policy,138*, 111232.

Lin, T.-P., Lin, F.-Y., Wu, P., Hämmerle, M., Höfle, B., Bechtold, S., Hwang, R., & Chen, Y. (2017). Multiscale analysis and reduction measures of urban carbon dioxide budget based on building energy consumption. *Energy and Buildings, 153*, 356–367. https://doi.org/10.1016/j.enbuild.2017.07.084

Lomas, K. J. (2010). Carbon reduction in existing buildings: a transdisciplinary approach. *Building Research & Information, 38*(1: Carbon reduction in existing buildings), pp. 1–11.

Lyu, X., Shi, A., & Wang, X. (2020). Research on the impact of carbon emission trading system on low-carbon technology innovation. *Carbon Management, 11*(2), 183–193.

Ma, H., Oxley, L., & Gibson, J. (2009). China's energy situation in the new millennium. *Renewable and Sustainable Energy Reviews, 13*(8), 1781–1799.

Mabrouk, S. B., Mabrouk, A. B., Harzli, K., La Cascia, D., et al. (2015). Experimentation of sustainable energy microsystems: The DE. DU. ENER. T. research project. In *2015 International Conference on Renewable Energy Research and Applications (ICRERA)* (pp. 1659–1664). IEEE, November 2015.

Martino, A., & e Territorio, T. T. (2011). Milano: From pollution charge to congestion charge. *Presentation to TML Leuven, 5*.

Mathur, S., Kaul, S., Gruteser, M., & Trappe, W. (2009, May). ParkNet: A mobile sensor network for harvesting real time vehicular parking information. In *Proceedings of the 2009 MobiHoc S3 workshop on MobiHoc S3* (pp. 25–28).

Matthäus, D. (2020). Designing effective auctions for renewable energy support. *Energy Policy, 142*, 111462.

Menendez-Carbo, S., Ruano, M. A., & Zambrano-Monserrate, M. A. (2020). Tourism Management Perspectives. *Tourism Management, 36*, 100727.

Mercado, R. G. (1998). *Environment and natural resources management: Lessons from city program innovations* (No. 1998-32). PIDS Discussion Paper Series.

Mercure, J. F., Knobloch, F., Pollitt, H., Paroussos, L., Scrieciu, S. S., & Lewney, R. (2019). Modelling innovation and the macroeconomics of low-carbon transitions: Theory, perspectives and practical use. *Climate Policy, 19*(8), 1019–1037.

Metz, D. (2018). Tackling urban traffic congestion: The experience of London, Stockholm and Singapore. *Case Studies on Transport Policy, 6*(4), 494–498.

Miller, G. (2020). Beyond 100% renewable: Policy and practical pathways to 24/7 renewable energy procurement. *The Electricity Journal, 33*(2), 106695.

Millward, H. (2006). Urban containment strategies: A case-study appraisal of plans and policies in Japanese, British, and Canadian cities. *Land Use Policy, 23*(4), 473–485.

Moon, S. Y., Jang, D., Kim, H. S., Lee, J. Y., & Kim, J. (2020). Importance of government roles for market expansion of eco-village development plan establishment research: Case study in the city of Suwon, South Korea. *Sustainability, 12*(24), 10293.

Mussone, L. (2017, June). Analysis of entering flows in the congestion pricing area C of Milan. In *2017 IEEE International Conference on Environment and Electrical Engineering and 2017 IEEE Industrial and Commercial Power Systems Europe (EEEIC/I&CPS Europe)* (pp. 1–6). IEEE.

Nelson, A., Prabakar, K., Nagarajan, A., Nepal, S., Hoke, A., Asano, M., Ueda, R., & Ifuku, E. (2017, April). Power hardware-in-the-loop evaluation of PV inverter grid support on Hawaiian electric feeders. In *2017 IEEE Power & Energy Society Innovative Smart Grid Technologies Conference (ISGT)* (pp. 1–5). IEEE.

Newton, P. W. (2010). Beyond greenfield and brownfield: The challenge of regenerating Australia's greyfield suburbs. *Built Environment, 36*(1), 81–104.

Nouri, H., Chavoshi Borujeni, S., & Hoekstra, A. Y. (2019). The blue water footprint of urban green spaces: An example for Adelaide, Australia. *Landscape and Urban Planning, 190*, 103613.

Nouri, H., Chavoshi Borujeni, S., & Hoekstra, A. (2020). Evapotranspiration, water demand and water footprint of urban green spaces. In *The 22nd EGU General Assembly Conference Abstracts* (p. 5512), held online 4–8 May 2020. Available from https://ui.adsabs.harvard.edu/abs/2020EGUGA..22.5512N/abstract.

Okata, J., & Murayama, A. (2011). Tokyo's Urban Growth, Urban Form and Sustainability. In: *Sorensen, A., & Okata, J. (eds.). Megacities: Urban Form, Governance, and Sustainability. Library for Sustainable Urban Regeneration*, vol 10. Springer, Tokyo. https://doi.org/10.1007/978-4-431-99267-7_2

Parks, D. (2020). Promises and techno-politics: Renewable energy and Malmö's vision of a climate-smart city. *Science as Culture, 29*(3), 388–409.

Pastorello, C., Dilara, P., & Martini, G. (2011). Effect of a change towards compressed natural gas vehicles on the emissions of the Milan waste collection fleet. *Transportation Research Part D: Transport and Environment, 16*(2), 121–128.

Peng, W., Baghzouz, Y., & Haddad, S. (2013, June). Local load power factor correction by grid-interactive PV inverters. In *2013 IEEE Grenoble conference* (pp. 1–6). IEEE.

Percoco, M. (2013). Is road pricing effective in abating pollution? Evidence from Milan. *Transportation Research Part D: Transport and Environment, 25*, 112–118.

Percoco, M. (2014). The effect of road pricing on traffic composition: Evidence from a natural experiment in Milan, Italy. *Transport Policy, 31*, 55–60.

Perrotti, D., & Iuorio, O. (2019). Green infrastructure in the space of flows: An Urban metabolism approach to Bridge environmental performance and user's wellbeing. In *Planning cities with nature* (pp. 265–277). Springer.

Pierce, G., & Shoup, D. (2013). Getting the prices right: An evaluation of pricing parking by demand in San Francisco. *Journal of the American Planning Association, 79*(1), 67–81.

Poggi, F., Firmino, A., & Amado, M. (2017). Assessing energy performances: A step toward energy efficiency at the municipal level. *Sustainable Cities and Society, 33*, 57–69.

Polzin, F. (2017). Mobilizing private finance for low-carbon innovation—A systematic review of barriers and solutions. *Renewable and Sustainable Energy Reviews, 77*, 525–535.

Qu, Y., & Liu, Y. (2017). Evaluating the low-carbon development of urban China. *Environment, Development and Sustainability, 19*(3), 939–953.

Rahman, S. M. R., Kabir, E., & Mannan, M. R. (2015). A scoping review of congestion pricing: Past work and its implementation benefits. *Port City International University Journal, 1851120791*(01773225500), 74.

Raven, R., Sengers, F., Spaeth, P., Xie, L., Cheshmehzangi, A., & de Jong, M. (2019). Urban experimentation and institutional arrangements. *European Planning Studies, 27*(2), 258–281.

Razzaghmanesh, M., Beecham, S., & Salemi, T. (2016). The role of green roofs in mitigating urban heat Island effects in the metropolitan area of Adelaide, South Australia. *Urban Forestry & Urban Greening, 15*, 89–102.

Rodier, C. J., & Shaheen, S. A. (2010). Transit-based smart parking: An evaluation of the San Francisco bay area field test. *Transportation Research Part C: Emerging Technologies, 18*(2), 225–233.

Roy, J., Ghosh, D., Ghosh, A., & Dasgupta, S. (2013). Fiscal instruments: Crucial role in financing low carbon transition in energy systems. *Current Opinion in Environmental Sustainability, 5*(2), 261–269.

Ruprecht, A. A., & Invernizzi, G. (2009). L'effetto del provvedimento di restrizione del traffico nel centro di Milano (Ecopass) sull'inquinamento urbano da polveri: i risultati di uno studio pilota. *Epidemiologia & Prevenzione, 33*(1–2), 21–26.

San, S. (n.d.). Neighborhood of Korean City Suwon goes car-free. Available from https://newcities.org/haenggung-dong-suwon-korea-car-free-zone/.

Santos, G. (2005). Urban congestion charging: A comparison between London and Singapore. *Transport Reviews, 25*(5), 511–534.

Schmid, E., & Knopf, B. (2012). Ambitious mitigation scenarios for Germany: A participatory approach. *Energy Policy, 51*, 662–672.

Scott, D., Gössling, S., Hall, C. M., & Peeters, P. (2016). Can tourism be part of the decarbonized global economy? The costs and risks of alternate carbon reduction policy pathways. *Journal of Sustainable Tourism, 24*(1), 52–72.

Sgobbo, A., & Basile, M. (2017). Sharing sustainability. *UPLanD-Journal of Urban Planning, Landscape & Environmental Design, 2*(2), 255–297.
Shang, T., Yang, L., Liu, P., Shang, K., & Zhang, Y. (2020). Financing mode of energy performance contracting projects with carbon emissions reduction potential and carbon emissions ratings. *Energy Policy, 144*, 111632.
Sharifi, E., Sivam, A., & Boland, J. (2016). Resilience to heat in public space: A case study of Adelaide, South Australia. *Journal of Environmental Planning and Management, 59*(10), 1833–1854.
Sharma, V., & Chandel, S. S. (2013). Performance analysis of a 190 kWp grid interactive solar photovoltaic power plant in India. *Energy, 55*, 476–485.
Shi, Q., & Lai, X. (2013). Identifying the underpin of green and low carbon technology innovation research: A literature review from 1994 to 2010. *Technological Forecasting and Social Change, 80*(5), 839–864.
Sorrentino, M., & Passerini, K. (2010). Evaluating public programs implementation: An exploratory case study. *International Journal of Electronic Government Research (IJEGR), 6*(3), 1–13.
Strachan, N., & Kannan, R. (2008). Hybrid modelling of long-term carbon reduction scenarios for the UK. *Energy Economics, 30*(6), 2947–2963.
Stripple, J., & Bulkeley, H. (2019). Towards a material politics of socio-technical transitions: Navigating decarbonisation pathways in Malmö. *Political Geography, 72*, 52–63.
Sukhwani, V., Shaw, R., Deshkar, S., Mitra, B. K., & Yan, W. (2020). Role of smart cities in optimizing water-energy-food Nexus: Opportunities in Nagpur, India. *Smart Cities, 3*(4), 1266–1292.
Sun, J., Cheshmehzangi, A., & Wang, S. (2020). Green infrastructure practice and a sustainability key performance indicators framework for neighbourhood-level construction of sponge city programme. *Journal of Environmental Protection, 11*(2), 82–109.
Suwon. (2013). EcoMobility Impulse, ICLEI—Local Governments for Sustainability, ICLEI Report.
Tian, X., Dai, H., Geng, Y., Huang, Z., Masui, T., & Fujita, T. (2017). The effects of carbon reduction on sectoral competitiveness in China: A case of Shanghai. *Applied Energy, 197*, 270–278.
Tiesdell, S. (2008). *Brownfields redevelopment and the quest for sustainability*. Emerald Group Publishing.
Tiwari, A. K., & Kalamkar, V. R. (2016). Performance investigations of solar water pumping system using helical pump under the outdoor condition of Nagpur, India. *Renewable Energy, 97*, 737–745.
Trencher, G., & Van der Heijden, J. (2019). Instrument interactions and relationships in policy mixes: Achieving complementarity in building energy efficiency policies in New York, Sydney and Tokyo. *Energy Research & Social Science, 54*, 34–45.
Trivellato, B., Mariani, L., Martini, M., & Cavenago, D. (2019). Leading knowledge mobilization for public value: The case of the congestion charge zone (Area C) in Milan. *Public Administration, 97*(2), 311–324.
Tyfield, D., Ely, A., & Geall, S. (2015). Low carbon innovation in China: From overlooked opportunities and challenges to transitions in power relations and practices. *Sustainable Development, 23*(4), 206–216.
van der Jagt, A. P., Smith, M., Ambrose-Oji, B., Konijnendijk, C. C., Giannico, V., Haase, D., Lafortezza, R., Nastran, M., Pintar, M., Železnikar, Š., & Cvejić, R. (2019). Co-creating urban green infrastructure connecting people and nature: A guiding framework and approach. *Journal of Environmental Management, 233*, 757–767
Wang, N., & Chang, Y. C. (2014). The evolution of low-carbon development strategies in China. *Energy, 68*, 61–70.
Werulkar, A. S., & Kulkarni, P. S. (2015). A case study of residential solar photovoltaic system with utility backup in Nagpur, India. *Renewable and Sustainable Energy Reviews, 52*, 1809–1822.
Wu, W. L. (2014). Veggie Power. *Tzu Chi Medical Care, 17*, 22–29.
Xi, J. (2017, October). Secure a decisive victory in building a moderately prosperous society in all respects and strive for the great success of socialism with Chinese characteristics for a new era.

In *Delivered at the 19th National Congress of the Communist Party of China October* (Vol. 18, No. 2017).
Xu, B., Wolfson, O., Yang, J., Stenneth, L., Philip, S. Y., & Nelson, P. C. (2013, June). Real-time street parking availability estimation. In *2013 IEEE 14th International Conference on Mobile Data Management* (Vol. 1, pp. 16–25). IEEE.
Yin, H., Zhao, J., Xi, X., & Zhang, Y. (2019). Evolution of regional low-carbon innovation systems with sustainable development: An empirical study with big-data. *Journal of Cleaner Production, 209*, 1545–1563.
Yu, S., Hu, X., Li, L., & Chen, H. (2020). Does the development of renewable energy promote carbon reduction? Evidence from Chinese provinces. *Journal of Environmental Management, 268*, 110634.
Yuan, H., Zhou, P., & Zhou, D. (2011). What is low-carbon development? A conceptual analysis. *Energy Procedia, 5*, 1706–1712.
Zhan, C., & de Jong, M. (2018). Financing eco cities and low carbon cities: The case of Shenzhen International Low Carbon City. *Journal of Cleaner Production, 180*, 116–125.
Zhang, D., Wang, J., Lin, Y., Si, Y., et al. (2017). Present situation and future prospect of renewable energy in China. *Renewable and Sustainable Energy Reviews, 76*, 865–871.
Zhang, M. M., Wang, Q., Zhou, D., & Ding, H. (2019). Evaluating uncertain investment decisions in low-carbon transition toward renewable energy. *Applied Energy, 240*, 1049–1060.
Zhang, S., Andrews-Speed, P., & Ji, M. (2014). The erratic path of the low-carbon transition in China: Evolution of solar PV policy. *Energy Policy, 67*, 903–912.
Zhang, T. (2019). Which policy is more effective, carbon reduction in all industries or in high energy-consuming industries? From dual perspectives of welfare effects and economic effects, *Journal of Cleaner Production, 216*, 184–196.
Zhao, G., Guerrero, J. M., Jiang, K., & Chen, S. (2017). Energy modelling towards low carbon development of Beijing in 2030. *Energy, 121*, 107–113.
Zheng, Y., Rajasegarar, S., & Leckie, C. (2015, April). Parking availability prediction for sensor-enabled car parks in smart cities. In *2015 IEEE Tenth International Conference on Intelligent Sensors, Sensor Networks and Information Processing (ISSNIP)* (pp. 1–6). IEEE.
Zhou, Y., Hao, F., Meng, W., & Fu, J. (2014). Scenario analysis of energy-based low-carbon development in China. *Journal of Environmental Sciences, 26*(8), 1631–1640.

Chapter 10
Evaluating China's Recent Low Carbon Progress

10.1 China's Progress of Low Carbon Development

In this chapter, we briefly evaluate China's current status of low carbon development, and specifically by looking into its recent progress. While some of the details of recent progress are already covered in previous chapters, in this chapter we highlight four key areas of 'energy consumption', 'renewable energy', 'Energy efficiency (buildings and construction)', and 'smart grids'. We then summarise some key challenges before exploring China's next steps in achieving sustainability transitions.

As discussed in the earlier chapters, the 13th Five-Year Plan (FYP) sets forth China's strategic intentions and defines its significant objectives, tasks, and measures for economic and social development. This plan is to serve as a guide to action for market entities, an important basis for government in performing its duties, and a shared vision to be shared among the people of China (Cheshmehzangi et al. 2018a). To achieve the goals of the 13th FYP, the relevant national departments respectively formulated various special plans such as "the 13th FYP for the Development of an Integrated Modern Transportation System", "the 13th FYP for Energy Development", "the 13th FYP for the Development of Renewable Energy", "the 13th FYP for the development of the construction industry", and so on (ibid).

Here, our focus is on China's more recent progress on low carbon and climate-resilient development. The progress is important for the next steps of China's plan for achieving the 2030 targets and the 2060 goals (see Chap. 11). To obtain a reasonable progress analysis, we focus on a detailed review of the aims and implantation progress of the 13th five-year plan. In doing so, we also aim to identify bottlenecks and opportunities for actions in this challenging decade of 2020–2030. The following sections explore four areas, as mentioned before. Here, we suggest some challenges as well before we conclude the chapter. For each area, we explore three aspects of 'goals and tasks', 'policies and action', and 'progress and effects'.

A. Cheshmehzangi and H. Chen, *China's Sustainability Transitions*,
https://doi.org/10.1007/978-981-16-2621-0_10

10.2 Energy Consumption

10.2.1 Goals and Tasks

In the field of energy consumption, the 13th FYP proposed the goal of promoting the energy consumption revolution and controlling total energy consumption within 5 billion tons of standard coal. The "13th FYP for Energy Development", "Strengthening Climate Change Response—China's National Independent Contribution", "13th FYP for Controlling Greenhouse Gas Emissions", and other special programs proposed two major objectives on the basis of the 13th FYP. Firstly, in terms of total energy consumption and intensity in 2020, total energy consumption was targeted to be controlled within 5 billion tons of standard coal, total coal consumption within 4.1 billion tons, and energy consumption per unit of GDP to fall by 15% compared to 2015 figures. Secondly, in terms of energy consumption structure in 2020, the proportion of non-fossil energy consumption was set to increase to over 15%, natural gas consumption ratio, which was aimed to reach 10%. The proportion of coal consumption was set to fall below 58%, and the proportion of coal used for power generation should account for more than 55% of coal consumption. While some of the goals were achieved, China's increasing energy consumption is recognised to be alarming in this current decade, i.e., throughout the lifetime of the 14th and 15th FYPs.

10.2.2 Policies and Action

In terms of controlling energy consumption and energy intensity, total energy consumption and energy consumption are regarded as important indicators of economic and social development. The People's Republic of China (PRC) implemented "double control" of total energy consumption and intensity (Ding et al. 2019; Yan & Su, 2020; Zhu et al. 2018). It also implemented a target responsibility system, decomposed the national "double control" target to all regions, major industries, and key energy-using units, and strengthened target responsibility evaluation. Each region clearly defined the annual work targets according to the tasks assigned by the state and resolved them at various levels. Besides, three sets of coal substitution policies were implemented (Niu et al. 2017; Shou et al. 2020). The first step was to control the total amount of coal consumption strictly. This step then led to the issuance of the "Circular on Strictly Controlling the Requirements for Planning and Construction of Coal-Fired Power Generation Projects in Key Areas". By the end of 2015, many provinces have issued their respective implementation plans for total consumption control and target responsibility management. The second was aimed at the reduction of substitution. In July 2016, the National Development Reform Committee (NDRC) issued the "Notice on Doing a Good Job in 2016 Coal Consumption Reduction Substitution" (Li & Lin, 2016; Cheshmehzangi et al. 2018a). The scope of coal substitution and reduction work included "Yangtze River Delta,

Beijing-Tianjin-Hebei, and the Pearl River Delta" and three provinces of Liaoning, Shandong, and Henan. The third policy area focused on collaborative governance (Deng et al. 2020). The Development and Reform Commissions of Beijing, Tianjin, and Hebei jointly issued the "Beijing-Tianjin-Hebei Energy Cooperative Development Action Plan (2017–2020)" to promote coal reduction. This could be a significant starting point for coal substitution plans.

In terms of expanding the natural gas consumer market, the natural gas price reforms were promoted (Paltsev & Zhang, 2015; Cheshmehzangi et al. 2018a). In recent years, the natural gas pipeline transmission price supervision system framework has been completed. In this process, three steps were considered. The first step was to establish a cost and income constraint mechanism for the price of gas distribution and to make restrictive provisions on the core index parameters that directly affect the gas distribution price. The second was to establish an incentive mechanism. The profits, which are from the gas company's own efforts to make the actual cost lower than the benchmark cost, are shared by gas companies and users. The third was to promote corporate information openness and require gas companies to actively disclose the price, cost, and other relevant information. Furthermore, the production and life energy use modes are innovated. The PRC implements clean energy-saving actions (Zhao et al. 2019) such as industrial energy conservation, green construction, and green transportation (Cheshmehzangi 2016b; Cheshmehzangi et al. 2018a). In addition, the PRC improved the energy conservation standard system (Yuan et al. 2017), vigorously developed and promoted energy-saving and high-efficiency technologies and products. In doing so, it also achieved full coverage of key energy-using industries and equipment energy-saving standards. The energy efficiency "leader" system is identified as the critical energy use industry (Cheshmehzangi et al. 2018a). It is also the system for benchmarking and compliance assessment, which are implemented in recent years.

10.2.3 Progress and Effects

During the 13th FYP, energy consumption indicators have achieved a satisfactory degree. In 2017, the total energy consumption was 4.49 billion tons of standard coal. At the same time, energy consumption per unit of GDP was reduced by 3.7%, and the unit of GDP of carbon dioxide emissions was reduced by 5.1%. The proportion of non-fossil energy consumption rose to 13.8%, the proportion of natural gas consumption reached 7%, and the proportion of coal consumption reduced to 60%. More importantly, the 2020 targets were mostly met as well (Table 10.1).

Table 10.1 Progress on key indicator for energy consumption

Indicator	2015	2020	Annual average increase rate [cumulative]	2016	2017	Implementation progress
Total energy consumption (100 Mtce)	43.4	<50 or 49.8	<3 or 2.9	44.1	45.6	Only increased approx. 25–28% of the incremental target Better than targeted
Reduction in energy consumption per unit of GDP (%)	/	15	3.2	5	3.7	Completed 57% of the decline target Better than targeted
CO_2 emissions reduction per unit of GDP (%)	/	18	3.9	6.6	5.1	Completed 63% of the decline target Better than targeted
Total coal consumption (100 million tons of raw coal)	39.6	41	0.70%	37.8	38.6	Total coal consumption declined. Better than targeted
Coal control - proportion in total primary energy (%)	63.8	56.8	[−6]	62.2	60.6	Faster than targeted
Non-fossil Energy share in primary energy consumption (%)	12	15	[3]	13.3	13.8	60% Faster than target
The share of natural gas in total energy consumption (%)	5.8	10	[4.1]	6.1	6.9	24% in 2016, faster than target

Source The Authors, developed based on the original National data extracted from https://data.stats.gov.cn/

10.3 Renewable Energy

10.3.1 Goals and Tasks

In the field of renewable energy, the 13th FYP proposed the goal of building a modern energy system. The main documents included "The 13th FYP for Energy Development", "13th FYP for Renewable Energy Development", "13th FYP for Solar Energy Development", "13th FYP for Wind Power Development", "Enhanced Actions on Climate Change: China's Intended Nationally Determined Contributions" and "13th FYP for Controlling Greenhouse Gas Emissions". In addition, other special programs proposed six primary objectives based on the 13th FYP (Cheshmehzangi et al. 2018a). We cover them in the following.

The first objective was to increase the total amount of renewable energy by the end of the 13th FYP. The first step was to increase the RE capacity. The annual utilisation of all renewable energy was proposed to be 730 million tons of standard coal. Among them, the commercial use of renewable energy was expected at 580 million tons of standard coal. The share of total renewable energy in primary energy consumption aimed to reach 12.5% by 2020. The 2020 data indicated that a higher percentage was reached by the end of the recent FYP. However, with almost four times higher coal consumption between 1990 and 2018 (IEA World Energy Balances, 2020), we see significant challenges in balancing energy supply by share and consumption changes. Table 10.2 below highlights some of the 13th FYP interim changes regarding primary energy supply. The data covers the timeline of two recent FYPs (i.e., the beginning of the 12th and mid-term of the 13th FYPs). To date, coal energy supply is slightly lowered to just below 57%, and nuclear energy values remain at 2%.

Furthermore, as shown in Fig. 10.1, from 2010 to 2020, the proportion of coal consumption in China's total energy consumption dropped from 69.2 to 56.7%, a decrease of 12.5 percentage points. The proportion of clean energy consumption in total energy consumption increased from 13.4 to 24.4%, an increase of 11 percentage points (Chnergy Institution, 2021). The energy consumption structure continued to shift towards green and low carbon. Therefore, the recent decade, including two recent FYPs, demonstrate tangible progress in low carbon transitions.

The second objective was to increase the generation of renewable energy power by 2020. The total installed capacity of renewable energy power generation was estimated to be 68,000 kW, and the power generation at 1.9 trillion kWh, which could account for 27% of the total power generation. However, the leading development was on capacity development rather than immediate power generation. By the end of the 13th FYP, the annual power generations of hydropower, wind power, photovoltaic, solar thermal, and biomass were proposed to reach 12,500, 4200, 1245, 200, and 90 billion kilowatts, respectively. The third objective was to increase renewable energy heating and fuel use by 2020. Renewable energy heating and civilian fuels collectively were proposed to replace approximately 150 million tons of standard coal fossil energy. Solar water heaters, geothermal energy utilisation, and biomass energy heating were planned to be 9600, 4000, and 150 billion tons of standard coal, respectively. The fourth objective was to increase the price competitiveness of wind power and photovoltaic projects by the end of the 13th FYP. The wind power project price was planned to compete with local coal-fired power generation platforms. PV project electricity prices could also be equivalent to the grid sales price, and the electricity price level could decline by more than 50% compared to the figures of 2015. The fifth objective was to solve the issue of renewable energy grid-connected operation and consumption by 2020. The total installed capacity of wind power grids in China was guaranteed to reach more than 210 million kilowatts. The installed capacity of grid-connected solar power was guaranteed to exceed 110 million kilowatts. Combined with the reform of the electricity market, the primary

Table 10.2 China Energy Supply by Share, including China's primary energy supply in 2010 and 2017, calculated using available data from IEA reports (IEA, 2021) and World Energy Data (2019)

Year	Fossil fuels (%)	Non-hyrdo renewables (%)	Coal (%)	Oil (%)	Gas (%)	Biofuels & waste (%)	Hydro (%)	Nuclear (%)	Geothermal (%)	Solar thermal (%)	Solar PV (%)	Marine	Wind (%)	Total (%)
2010	87.3	0.9	68	16	3	5	6	1	0.16	0.37	0.01	–	0.38	100
2017	82.6	4.1	59	17	6	3	8	2	0.37	0.82	0.89	–	2.02	100

Dashes indicate negligible or zero values

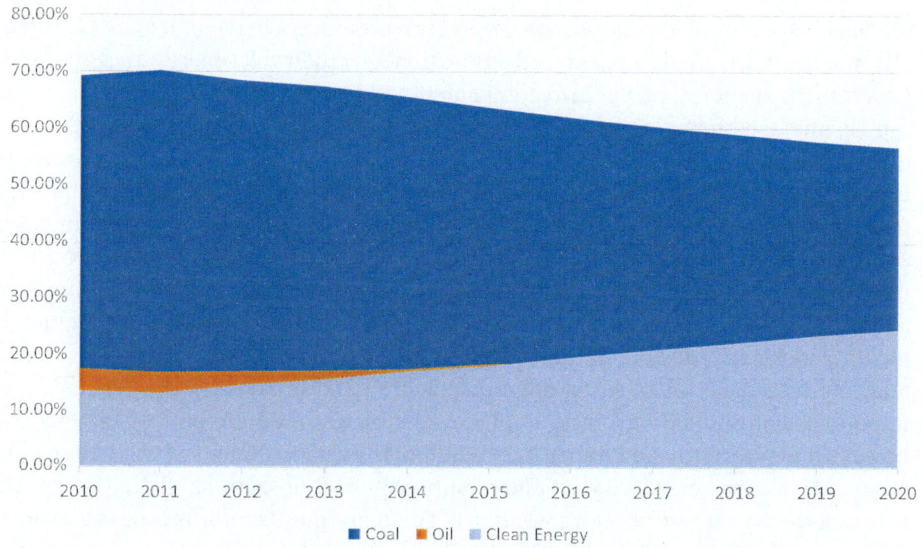

Fig. 10.1 Recent progress on coal, oil, and clean energy proportions of energy consumption in China from 2010 to 2020 (Developed and drawn by the authors using available data from Chnergy Institution, 2021)

aim was to resolve the issue of water and electricity abandonment. The annual utilisation hours of wind power and solar power in the power-restricted regions were planned to meet the requirements for full scale acquisitions. Lastly, the sixth objective was to improve the assessment and restriction mechanism for renewable energy indicators by 2020. In the 13th FYP, China planned to establish an indicator management system for the proportion of renewable energy in all provinces' total primary energy consumption (i.e., autonomous regions and municipalities directly under the Central Government) and the consumption of renewable energy in the entire society. In return, the proportion of non-hydro renewable energy generating capacity and coal-fired power generation of each power generation enterprise was expected to increase during the last five years (Cheshmehzangi et al. 2018a).

10.3.2 Policies and Action

In four areas, we see tangible progress during the 13th FYP. The first is the introduction of new subsidy systems, including the special subsidy policy for the renewable energy industry, the exclusive support policy for the wind power industry, and the exclusive support policy for the solar power industry. The second is the establishment of a target guidance mechanism. On the one hand, it decomposes the index of gravity by the annual implementation, which has a certain restrictive effect on increasing the installed renewable energy capacity in each province. On the other hand, it annually monitors the development and utilisation of renewable energy by provincial government agencies, power grid companies, and power generation companies. The third is the establishment of a fully guaranteed acquisition system. The "Management Measures for Total Entitlement of Renewable Energy Power Generation" was issued. The annual power generation capacity of renewable energy grid-connected power generation projects is divided into the portion of guaranteed power purchases and the portion of market transaction power. This system defines the responsibilities and obligations of the relevant market entities in implementing the safeguard acquisition. It also provides quantifiable benchmarks for implementing legal requirements and clearly describes the "preferential power generation dispatch" benchmarking boundary for power system dispatching and operation. This system is an important institutional design to implement the latest requirements of the law and power system reform. And fourth, we highlight the establishment of two markets for renewable energy, electricity voluntary subscription transactions and green certificate mandatory subscription transactions. The main purpose for these is to address the issue of funding gaps in electricity tariff subsidies and guide the green electricity consumption of the entire society to manage renewable energy development issues (Cheshmehzangi et al. 2018a).

10.3.3 Progress and Effects

During the 13th FYP, the renewable energy sector's main objectives have achieved a good degree, and most of the indicators have completed more than 50% of their expectations (Table 10.3). We summarise four key progress in the following:

(1) The total amount of renewable energy has increased. As early as 2016, the total annual renewable energy use reached 558 million tons of standard coal;
(2) The generation of renewable energy continues to grow;
(3) With the improvement of renewable energy technologies, the price of electricity for wind power and photovoltaic projects has dropped;
(4) The capacity of renewable energy grid operation and consumption has been improved (Cheshmehzangi 2018).

10.4 Energy Efficiency (Buildings and Construction)

10.4.1 Goals and Tasks

In the field of energy consumption, the 13th FYP proposed the goal of promoting building energy efficiency and achieving low carbon development in the building sector. Primary programs such as "The 13th FYP for Energy Development", "The

Table 10.3 Progress on key indicator for renewable energy. Dashes indicate no available information or data in specific years

Indicator	2015	2016	2017	2020	Implementation progress
Annual use of all renewable energy (100 million tons of standard coal)	5.2	5.8	6.2	7.3	–
Non-fossil energy share in primary energy consumption (%)	12.1	13.3	13.6	15	60% Faster than target
Total renewable energy power generation (100 GW)	4.8	5.7	6.5	6.8	85% Faster than target
Total renewable energy generation (trillion kW)	1.38	–	1.7	1.9	62% Faster than target
Total renewable energy in total generation (%)	25	25.7	26.4	27	70% Faster than target
Hydropower installation (100 GW)	3.2	3.3	3.41	3.4	Target reached Expected to be better than target
Grid-connected wind power installation (100 million kW)	1.29	1.49	1.64	2.1	43% Faster than target
Grid-connected photovoltaic installation (100 million kW)	0.43	0.77	1.3	1.05	Target reached Expected to be better than target
Hydropower generation (100 million kWh)	9960	11,807	11,945	12,500	78% Faster than target
Wind power generation (100 million kWh)	1863	2410	3057	4200	51% Faster than target
Photovoltaic power generation (100 million kWh)	392	–	1182	1245	93% Faster than target

13th FYP for Construction Industry Development", "The 13th FYP for Building Energy Efficiency and Green Building Development", "Comprehensive Work Plan for Energy Saving and Emission Reduction during the 13th FYP period", "National Energy Conservation Action Plan during the 13th FYP period", "Enhanced Actions on Climate Change: China's Intended Nationally Determined Contributions", "The 13th FYP for Controlling Greenhouse Gas Emissions", and other special programs proposed five significant objectives in achieving low carbon development. First, the energy efficiency of newly built buildings in cities and towns was targeted to increase by 20% compared to 2015. In achieving this, energy-saving standards for key parts of buildings (e.g., for the envelope, insulation, doors, windows, etc.) were proposed to reach or approach the advanced international levels of the current time. Second, the proportion of green building areas in newly built urban buildings is proposed to exceed 50%, and that of green building materials exceeds 40%. This is regarded as a major move towards green building certification, assessment, and recognition, which would be utilised to construct newly built urban buildings. Third, the energy-saving reconstruction of existing residential buildings was proposed to be completed with an area of over 500 million square meters. The area of the energy-saving reconstruction of public buildings was planned to be 100 million square meters by the end of the 13th FYP. Also, the proportion of energy-saving buildings in the existing urban residential buildings was targeted to be more than 60%. Fourth, the proportion of conventional energy consumption of urban renewable energy alternatives to civil buildings was targeted to exceed 6%. And fifth, breakthroughs were made in rural buildings' energy efficiency in economically developed areas and key development areas. In this regard, the proportion of energy-saving measures adopted was targeted to exceed 10% by the end of the 13th FYP (Cheshmehzangi 2020a; b). We note that at this point, official reports are yet to confirm the achievement of such targets in the recent low carbon progress.

10.4.2 Policies and Action

In recent years, in terms of improving building energy efficiency and implementation quality standards, a model for leading projects for benchmarking (regional) standards was established. This also helps develop new policies or reform the older ones towards more updated standards and higher-level benchmarks. Simultaneously, the green building standards policy was implemented to promote the quantity and quality of green buildings. In addition, in terms of building energy efficiency, the model of demonstration cities is set up. In this regard, market mechanisms such as the model of contract energy management and public-private partnership (PPP) opportunities were recently explored to promote existing building renovation. Lastly, in terms of promoting the application of renewable energy buildings, the policy of mandatory application of renewable energy buildings was introduced during the 13th FYP (Cheshmehzangi 2018).

10.4.3 Progress and Effects

In general, we can verify the main objectives in the field of building energy efficiency have achieved a satisfactory degree. The earlier and interim reports indicate success in the early stages of the 13th FYP, meaning that the major indicators have reached the proposed targets. In 2016, urban green buildings accounted for 29% of newly built buildings, 98.8% of new green building standards were implemented in urban areas, and 788.7 billion square meters of the energy-saving renovation of existing residential buildings were implemented. The cumulative area of the energy-saving renovation of urban public buildings was 1.526 billion square meters. Other indicators suggest the energy efficiency of new buildings in urban areas has increased. In 2016, the proportion of mandatory standards for implementing energy conservation

Table 10.4 Progress on key indicator for building sector

Indicator	2015	2016	2020	Implementation progress Evaluation
Percentage of green buildings of newly built buildings in the urban area (%)	20	29	50	30% in 2016 Better than targeted
Implementation of existing residential buildings energy-saving retrofit (billion square meters)	–	8,789	–	–
Implementation rate of urban newly built green building standards (%)	20	98.8	50	Much better than targeted
Urban public building energy-saving retrofit cumulative area (billion square meters)	12.5	15.26	17.5	55% in 2016 Better than targeted

Dashes indicate no available information or data in specific years

in newly built urban buildings nationwide was 98.8%. In addition, the proportion of green buildings in newly constructed buildings in urban areas has increased. As of the end of 2016, over 20,000 mandatory green building standard projects have been completed in China, and 7235 architectural projects have been awarded green building evaluation marks nationwide with a building area of more than 800 million square meters (Cheshmehzangi 2018). Since then, a significant shift has occurred, and more projects are awarded green building evaluation marks. More importantly, the local standards have entered the green assessment and certification competition more rigorously. In particular, we see a huge progression from 2017 onwards. In other areas, there is evidence that shows the area for the energy-saving transformation of existing residential buildings has been increased, and the energy-saving transformation of public buildings has been accelerated. The proportion of renewable energy substituting for the conventional energy consumption of civil buildings in urban areas has also increased (Dawodu & Cheshmehzangi, 2017; Cheshmehzangi 2018). Table 10.4 below summarises some of the key areas of progression on critical indicators for the building sector. These indicators include (1) percentage of green buildings of newly built buildings in the urban area, (2) implementation of existing buildings energy-saving retrofit (billion square meters), (3) implementation rate of urban newly-built green building standards, and (4) Urban public building energy-saving retrofit cumulative area (billion square meters).

10.5 Smart Grids

10.5.1 Goals and Tasks

In the field of smart grids, the 13th FYP proposed the goal of building a modern energy system. This has been part of the strategies of the 11th and 12th FYPs and the current 14th FYP. This goal is likely to be a central one until the ultimate goal of carbon neutrality is achieved successfully. Primary programmes, such as "The 13th FYP for Energy Development", "The 13th FYP for Electric Power Development", "Enhanced Actions on Climate Change: China's Intended Nationally Determined Contributions", "The 13th FYP for Controlling Greenhouse Gas Emissions", and other special programs proposed the objectives of upgrading the distribution network and promoting smart grid construction based on the 13th FYP (Cheshmehzangi 2018). In recent years, in particular, we see more advancement in smart technologies, smart grid systems, and modernisation of energy systems. Some regions have gone through major industrial restructuring, and some have already initiated the shift from high polluting to low-polluting productions.

10.5.2 Policies and Action

During the 13th FYP, the power distribution network was upgraded, the construction of smart grids was promoted, and the model of the demonstration project was established. The relevant policies issued during the 12th FYP still had a positive effect on the situation of the last five years. In May 2015, the National Energy Administration (NEA) issued the "Guiding Opinions on Promoting the Construction of a New Energy Micro-grid Demonstration Project" to explore the establishment of an integrated local area for transmission, distribution, and storage that accommodates high-volatility renewable energy. The guiding opinions were targeted to explore new commercial operation modes and new business models of electric energy services. Moreover, they were intended to promote more dynamic electricity market innovation and development and form a completely new energy micro-grid technology system and management system (Cheshmehzangi 2018). In particular, we see major shifts towards smaller scales of communities/neighbourhoods and districts (Cheshmehzangi et al. 2021), which included the recognition of smart grids and smart energy systems at smaller spatial levels.

10.5.3 Progress and Effects

During the 13th FYP, we notice the level of smart devices, intelligent smart energy meters, and electricity consumption information collection systems have gradually increased. These progressions motivated many cities and regions (as well as companies, businesses, local governments, and relevant stakeholders) to promote smart energy systems and technologies in achieving, or at least implementing the first steps towards, low carbon transitions (Cheshmehzangi 2016a). However, the negative aspect could be seen from the market point of view, where the shift has been mainly focused on economic growth and development of new businesses and high-tech industries. Nonetheless, the last five years' achievements cannot be ignored, especially that we see more popularity of smart systems at larger scale projects of communities and neighbourhoods. Also, the integration of smart systems at the building level is a major shift in regards to energy use optimisation, smarter building management systems, resource management in buildings, etc. The use of blockchain and other similar methods are good examples, which have become more popular during the timeline of the 13th FYP.

For example, in 2016, approximately 5000 intelligent substations of various voltage classes were built. In the same year and only one year after the start of the 13th FYP, 40 new generation smart substations featuring "integrated equipment, integrated networks, and integrated systems" have been put into operation. The shift has been phenomenal, as the emphasis has been increasing on smartness and smart integration for low carbon progression. Also, in 2016, six regional dispatch centers in North China, East China, Central China, Northeast China, Northwest China, and Southwest China, and 27 provincial-level dispatch centers in Beijing, Hebei, Jiangsu, and Hubei have all completed smart grid dispatch and control systems (Cheshmehzangi 2018). The five southern provinces and regions continued to promote the construction of an integrated power grid operation intelligence system and completed the OS/2 technical standard system (i.e., a series of computer operating systems), key technology research, and pilot construction work. By the end of 2016, newly installed smart electricity meters in 26 provinces and cities such as Beijing, Jiangsu, Hubei, etc., totaled 74.66 million and completed the construction of provincial-level electricity information collection system master stations. The accumulated electricity consumption information of 410 million households was collected and evaluated. For these projects, smart meter coverage in the five southern provinces reached 81%, which was a rapid increase in a relatively short time (ibid.). The more recent signs of progress include the integration of new technologies such as blockchain (Hasankhani et al. 2021), internet of things (IoT) (Bagdadee et al. 2020), smart metering (Wang et al. 2018), as well as enhanced security systems (Wang et al. 2021), consideration of energy internet (Xiao et al. 2020), artificial intelligence (AI) application in energy systems and optimisation procedures (Shi et al. 2020), and the utilisation of government-market dynamics in socio-technical energy transitions (Mah, 2020). More importantly, the increasing popularity of electric vehicles, new technologies, and renewable energy resources has become a significant turning point towards the development of hybrid smart-low carbon models and systems.

10.6 Recent Challenges of the 13th FYP and What Lies Ahead

Here, we summarise five recent challenges in China's low carbon development, those that were experienced during the 13th FYP and are likely to be continuing in the next five years or so. First, we highlight that total control targets' completion of energy consumption targets in the 13th FYP is still under greater pressure.

With the further improvement of the future's economic situation and the further increase in people's living energy demand, the total energy consumption growth rate may increase further. We also anticipate the country's total energy consumption control targets remain under considerable pressure. With the ongoing COVID-19 pandemic, we notice disruptions in specific areas or sectors. The impact is also identified on the economic situation and strategies that were set earlier. Simultaneously, in some areas, several high-energy-consuming projects have been introduced for steady growth in recent years. Local governments in some regions do not pay enough attention to "double control", and work measures have weakened. During the 13th FYP period, energy consumption increased rapidly. Therefore, the completion of "Double Control" targets, mainly the total energy consumption control target, is still facing a lot of pressure (Cheshmehzangi 2018). The increasing levels of consumption and energy demand remain significant challenges of the following two FYPs, at least until the end of 2030. Second, China faces a shortage of natural gas supply in winter. Firstly, natural gas demand grows substantially. Secondly, the natural gas supply declines unexpectedly. Thirdly, the implementation of natural gas "non-preserving people" measures is not in place, and the dual-track pricing system for civil gas and non-civil gas is adopted. Fourthly, over the years, the construction of gas storage facilities has been insufficient, and the capacity of the peak has been seriously inadequate (Cheshmehzangi 2018).

Third, there is insufficient policy support for the development of renewable energy. Firstly, subsidy arrears are profound, affecting the benefits of renewable energy companies, and to a certain extent, harm the country's credibility. Secondly, the renewable energy development model and layout are not reasonable. The power generation base is far from the eastern power load center, and it is not conducive to the consumption of renewable energy. Thirdly, it is difficult to eradicate the problem of renewable energy grid-connected operation and consumption. Fourthly, the new policy about photovoltaic power generation may negatively affect (Cheshmehzangi 2018).

Fourth, with the improvement of living standards, the demand for building energy continues to grow, resulting in a considerable potential for making energy consumption growth in the future. First, urban and rural residents' living standards have improved, and household appliances, energy-using equipment, and use time have all shown a clear growth trend, bringing about a continuous increase in residential energy consumption. In addition, inhabitants in the Yangtze River Basin are increasingly demanding heating, bringing about an increase in energy consumption and carbon emissions. Last but not least, smart grid development faces challenges. As a major aspect, big data processing is still facing challenges. The development time of smart grid big data processing is short. The development of big data processing lacks a scientific and reasonable management mechanism and lacks talents with solid professional knowledge and high professional qualities. As a result, the smart grid big data processing technology needs to be improved and cannot meet the need of increasing data processing volume. Besides, the smart grid investment is still under pressure (Cheshmehzangi 2018). In addition to the above five key challenges, the financial pressures apply to all five areas. Therefore, it could become even more challenging in the coming years.

10.7 Conclusions

If China discontinues its low carbon progress and continues with a business-as-usual model of development, then we anticipate significant growth in greenhouse gas emissions, energy demand, resource use, and climate impacts. While we still face significant challenges, we also appraise the progressive achievements of the last one or two FYPs. The perfect society model may still be far from reach, but the steps are already taken in sustaining the urbanisation progress, urban development, and low carbon futures. However, there are still issues with the pro-developer development mode and trendy unsustainable development projects (Cheshmehzangi, 2014) that have not yet changed the essence of urban development practices and urbanisation trends. China still faces growing energy demand trends and experiences societal modernisation processes, which remain as challenges and opportunities for the next steps of low carbon progression. Thus, we are yet to witness paradigm shifts, scaled-up development (based on earlier and ongoing experimentation projects) (Cheshmehzangi et al. 2018b; Raven et al. 2019), novel strategic practices, innovative progress, and policy reforms.

For us, the most important factor at this point is the rapid pace of progress in terms of low carbon and sustainable transitions. China has invested a lot in learning, experimenting, testing, and modelling what could become the next generation of urbanism and sustainable urban development models. Therefore, its next steps to achieve the carbon peak by 2030 and the longer-term plan of carbon neutrality by 2060 are crucial to how China could reverse its production trends, consumption behaviours, and energy sector. The next steps require step-by-step transitions, ensuring that low carbon targets and indicators are set to the highest levels and benchmarks are revisited and revised regularly.

In recent years, many Chinese reports highlight the interchangeabilities between urbanisation, climate change, and sustainability. The next steps of low carbon transitions are very much based on China's recent transitions. Therefore, this chapter analysed China's recent low carbon development progress by exploring explicit areas that are important to future policies and practices. This chapter, together with the next two chapters, help to identify pathways and sustainability transition directions. These will be very important within decarbonisation and sustainability transition areas (Baeumler et al. 2012; Rauland & Newman, 2015). They will be beneficial to the next steps of decarbonising cities (Butters et al. 2020), reducing carbon emissions, innovating new paradigms, and achieving the ultimate goals of carbon neutrality.

Acknowledgments Ali Cheshmehzangi acknowledges the Asian Development Bank (ADB) and the National Development Reform Commission (NDRC) of China for the opportunity of a research project that he led during 2017 and 2018. In particular, we acknowledge colleagues and friends from ADB for their support and the opportunity to review China's 13th Five-Year-Plan, which then identified the critical areas of low carbon development for the 14th Five-Year-Plan. Besides, we thank our team members at 'Chinese Society for Urban Studies (CSUS)' and 'National Center for Climate Change Strategy and International Cooperation (NCSC)', both based in Beijing, China. In particular, we acknowledge the support of Dr. Hailong LI (leading the CSUS Team) and Dr. Xiu YANG (leading the NCSC Team). With their support and valuable collaboration, we managed to put together a very detailed review and a comprehensive technical report.

References

Baeumler, A., Ijjasz-Vasquez, E., & Mehndiratta, S. (Eds.). (2012). *Sustainable low-carbon city development in China.* Washington DC: World Bank Publications.

Bagdadee, A. H., Hoque, M. Z., & Zhang, L. (2020). IoT based wireless sensor network for power quality control in smart grid. *Procedia Computer Science., 167*, 1148–1160.

Butters, C., Cheshmehzangi, A., & Sassi, P. (2020). Cities, energy and climate: Seven reasons to question the dense high-rise city. *Journal of Green Building, 15*(3), 197–214.

Cheshmehzangi, A. (2014). The urban and urbanism: China's new urbanization and approaches towards comprehensive development. *The International Journal of Interdisciplinary Environmental Studies, 8*(3–4), 1–12.

Cheshmehzangi, A. (2016a). Feasibility study of Songao's low carbon town planning, China. *Energy Procedia, 88*, 313–320.

Cheshmehzangi, A. (2016b). China's new-type urbanisation plan (NUP) and the foreseeing challenges for decarbonization of cities: A review. *Energy Procedia, 104*, 146–152.

Cheshmehzangi, A. (2018). The changing urban landscape of Chinese cities: Positive and negative impacts of urban design controls on contemporary urban housing. *Sustainability, 10*(8), 2839.

Cheshmehzangi, A. (2020a). Evaluating the Nexus between housing and energy sectors: The comparison of urban, Peri-urban and rural housing areas in Zhuhai, China. *Energy and Power Engineering, 12*(06), 314.

Cheshmehzangi, A. (2020b). Low carbon transition at the township level: Feasibility study of environmental pollutants and sustainable energy planning. *International Journal of Sustainable Energy*, 1–27. https://doi.org/10.1080/14786451.2020.1860042.

Cheshmehzangi, A., Dawodu, A., & Sharifi, A. (2021). *Sustainable urbanism in China.* Routledge.

Cheshmehzangi, A., Li, H., & Yang, X. (2018a). Low carbon and climate resilient urban development in China. In *ADB Report, Developed for the National Development Reform Committee (NDRC)*, Beijing, China, October 2018.

Cheshmehzangi, A., Xie, L., & Tan-Mullins, M. (2018b). The role of international actors in low-carbon transitions of Shenzhen's International Low Carbon City in China. *Cities, 74*, 64–74.

Chnergy Institution. (2021). The Energy Development Report of China 2021. <中国能源发展报告 2021>. Available from: http://www.chnergy.com/web/weixinbook/searchlistd1?caid=6.

Dawodu, A., & Cheshmehzangi, A. (2017). Impact of floor area ratio (FAR) on energy consumption at meso scale in China: Case study of Ningbo. *Energy Procedia, 105*, 3449–3455.

Deng, W., Cheshmehzangi, A., Ma, Y., & Peng, Z. (2020). Promoting sustainability through governance of eco-city indicators: A multi-spatial perspective. *International Journal of Low-Carbon Technologies.*

Ding, Q., Chen, H., Li, P., Liu, M., & Lin, L. (2019). Research on standard system for "double control" of total energy consumption and energy intensity. In *E3S Web of Conferences* (Vol. 118, p. 01020). EDP Sciences.

Hasankhani, A., Hakimi, S. M., Shafie-Khah, M., & Asadolahi, H. (2021). Blockchain technology in the future smart grids: A comprehensive review and frameworks. *International Journal of Electrical Power & Energy Systems, 129*, 106811.

IEA. (2021). *Data and statistics for China*, Available from: https://www.iea.org/data-and-statistics/data-tables?country=CHINA&energy=Balances&year=2017.

IEA World Energy Balances. (2020). *China's energy data*, Available from: https://www.iea.org/subscribe-to-data-services/world-energy-balances-and-statistics.

Li, J., & Lin, B. (2016). Inter-factor/inter-fuel substitution, carbon intensity, and energy-related CO_2 reduction: Empirical evidence from China. *Energy Economics, 56*, 483–494.

Mah, D. N. (2020). Conceptualising government-market dynamics in socio-technical energy transitions: A comparative case study of smart grid developments in China and Japan. *Geoforum, 108*, 148–168.

Niu, D. X., Song, Z. Y., & Xiao, X. L. (2017). Electric power substitution for coal in China: Status quo and SWOT analysis. *Renewable and Sustainable Energy Reviews, 70*, 610–622.

Paltsev, S., & Zhang, D. (2015). Natural gas pricing reform in China: Getting closer to a market system? *Energy Policy, 86*, 43–56.

Rauland, V., & Newman, P. (2015). *Decarbonising cities: Mainstreaming low carbon urban development*. Springer.

Raven, R., Sengers, F., Spaeth, P., Xie, L., Cheshmehzangi, A., & de Jong, M. (2019). Urban experimentation and institutional arrangements. *European Planning Studies, 27*(2), 258–281.

Shi, Z., Yao, W., Li, Z., Zeng, L., Zhao, Y., Zhang, R., Tang, Y., & Wen, J. (2020). Artificial intelligence techniques for stability analysis and control in smart grids: Methodologies, applications, challenges and future directions. *Applied Energy., 278*, 115733.

Shou, M. H., Wang, Z. X., Li, D. D., & Wang, Y. (2020). Assessment of the air pollution emission reduction effect of the coal substitution policy in China: An improved grey modelling approach. *Environmental Science and Pollution Research, 27*(27), 34357–34368.

Wang, Y., Qiu, H., Tu, Y., Liu, Q., Ding, Y., & Wang, W. (2018). A review of smart metering for future Chinese grids. *Energy Procedia, 152*, 1194–1199.

Wang, Z., Jiang, D., Wang, F., & Nowak, R. (2021). A polymorphic heterogeneous security architecture for edge-enabled smart grids. *Sustainable Cities and Society., 67*, 102661.

World Energy Data. (2019). *The energy system of the People's Republic of China*. Available from: https://www.worldenergydata.org/china/.

Xiao, Z., Wang, J., Yang, X., Xiao, Y., Liu, H., Wu, Y., Deng, K., & Chen, J. (2020). Research on the medium and long term development framework of smart grid under the background of energy Internet. *Procedia Computer Science, 175*, 769–773.

Yan, J., & Su, B. (2020). What drive the changes in China's energy consumption and intensity during 12th Five-Year Plan period? *Energy Policy, 140*, 111383.

Yuan, X., Zhang, X., Liang, J., Wang, Q., & Zuo, J. (2017). The development of building energy conservation in China: A review and critical assessment from the perspective of policy and institutional system. *Sustainability, 9*(9), 1654.

Zhao, X., Cheng, H., Zhao, H., Jiang, L., & Xue, B. (2019). Survey on the households' energy-saving behaviors and influencing factors in the rural loess hilly region of China. *Journal of Cleaner Production, 230*, 547–556.

Zhu, B., Jiang, M., Wang, K., Chevallier, J., Wang, P., & Wei, Y. M. (2018). On the road to China's 2020 carbon intensity target from the perspective of "double control." *Energy Policy, 119*, 377–387.

Chapter 11
Lessons and Paradigms to Meet the 2030 Targets and the 2060 Plan

11.1 A Brief Introduction

This chapter highlights some of the key lessons and paradigms for china's sustainability transition. We explicitly look into low carbon and climate-resilient practices and those that we could see critical for China's interim and ultimate steps in its pathways towards carbon neutrality. The interim step is regarded as the 2030 targets, and the ultimate step is the 2060 carbon-neutrality plan.

First, we provide ten suggestions in three primary areas: low carbon transport systems and infrastructures, urban form and planning, and renewable energies and smart grids (see Chap. 9). In doing so, we highlight key aspects of sustainable urban development in facing climate change (Cheshmehzangi & Dawodu, 2019) and contextual conditions of cities and urban sustainability ideals (Baeumler et al., 2012; Butters et al., 2020; Cheshmehzangi & Butters, 2017). Afterward, to complete the chapter, we suggest some key lessons and paradigms, which will be vital to meet the 2030 targets and the 2060 carbon-neutrality plan. We will provide ten low carbon and climate-resilient urban development paradigms in planning, policy reforms, low carbon strategies, smartness, sustainable urbanism, etc.

11.2 China's Next Steps in Achieving Sustainability Transitions

Here, we highlight ten suggestions that could be regarded as pathways towards China's 2030 targets and lessons for the 2060 carbon neutrality plan. These were initially discussed by Cheshmehzangi et al. (Cheshmehzangi, Li, et al., 2018) and are now narrated as part of China's next steps in achieving sustainability transitions, especially in low carbon development. These ten suggestions follow key recommendations of verticality of policy mixes for sustainability transitions (Huang, 2019), local responses to low carbon development (Bohnsack, 2018), and the importance

A. Cheshmehzangi and H. Chen, *China's Sustainability Transitions*,
https://doi.org/10.1007/978-981-16-2621-0_11

of horizontal and vertical linkages (Bai et al., 2009). We focus mainly on our three overarching areas of 'transportation', 'urban form and spatial planning', and 'renewable energies and smart grids'. The followings summarise these suggestions before we move into potential low carbon and climate-resilient paradigms.

(i) **Improve top-level design to guide collaborative planning and development**—Besides meeting the needs of the social and economic development (He et al., 2018; Yap & Truffer, 2019), as well as people's travel demands, China also requires the maximisation of the efficiency of resources and energy usage and the reduction of greenhouse gas (GHG) emissions. Thus, the current practice of treating low carbon planning (such as for transportation planning) solely as a specific plan should be changed to formulate an integrated urban development plan with clear target plans/guidance. By integrating low carbon development objectives into urban land planning, spatial planning, and economic planning, the city can avoid duplication of construction and ineffective construction and encourage compact city form that would contribute to the reduction of citizens' needs as well as better public transportation. These include responding to daily demands and implementing innovation intermediaries for integrated low carbon solutions.

(ii) **Promote structural, technical and managerial traffic energy conservation, and emission reduction**—Multimodal and integrated transportation should further optimise the structure of vehicle transportation capacity and improve the transportation network's operational efficiency. The integration of ICT in a multimodal transport (Harris et al., 2015) and sensory systems (Wang et al., 2020) are suitable examples that could be utilised for future progress in this area, as well as for logistics systems and meeting the environmental sustainability objectives (Rondineli & Berry, 2000), and development of integrated supply chain networks and infrastructures (Hajibabai & Ouyang, 2013). Priority should be given to the development of public transportation, which will significantly increase the share of public transport travel and future infrastructural development. Meanwhile, the city should explore innovative methods to maximise the road network's utilisation efficiency and increase the actual loading rate of passenger and cargo transportation. The direction towards a hybrid model of smart-low carbon development is an opportunity to optimise the operations, resources, and management systems.

(iii) **Develop smart technology for low carbon transportation**—Technologies like traffic information system (TIS) (Lin & Yang, 2002; Wang et al., 2016) has proved to be beneficial for transportation monitoring and management by providing supplicated on-road traffic condition information, which can further assist transportation improvement. Therefore, developing smart technology for low carbon transportation, such as the development of intelligence traffic guidance and control system and environmental monitoring technology, would provide great technical support for the overall development of low carbon transportation.

(iv) **Establish a low carbon supervision mechanism**—At present, China has formulated a series of policies to develop low carbon development, such as for low carbon transportation. However, there are no corresponding regulatory agencies and mechanisms to supervise them. By facilitating such mechanisms, we can measure and monitor low carbon city construction (Zhao et al., 2019) and evaluate low carbon construction-related standards (Xing, 2018). In doing so, we could enhance the low carbon economy's legal regulation through the mechanism of supervising agencies (Xie, 2014). Therefore, the regulatory responsibilities of the various agencies should be clearly defined in the relevant laws. Meanwhile, the government and industry should also establish and expand the low carbon development index system (Wang et al., 2021; Yang et al., 2018), assessment methods, and incentive and restraint mechanisms to provide transparent and standardised guidance low carbon development.

(v) **Drawing on the international experience and learn from advanced practices of low carbon development**—China, as a developing country with increasing environmental awareness and attention on developing a low carbon development, can benefit from the various experience and advanced practices in other countries. For example, in Germany, the government promulgated a new EMU tax system in 2009 to impose different taxes based on the car size. In Japan, a "green taxation" was implemented on the purchase of private cars in 2001. Complementing the stricter emission reduction regulations for automotive products, as an example, could lead to better sustainability transitions through gradual steps. For instance, the implementation of "green taxes" imposes increasingly stringent regulations on reducing automotive products. Learning from the experience of international low carbon development, China can explore low carbon fiscal policies and regulations such as "congestion fee", "car carbon tax", "new energy vehicle subsidy", "limit line", and "restriction purchase", which could fit in the Chinese context. Examples of governance by green taxes (Andersen, 1994) and relevant monitory mechanisms and protecting the environment (Oates, 1995) are already established as a practical approach to achieving low carbon transitions.

(vi) **Improving top-level and multidisciplinary cooperative design**—Smart grid is an open and widely connected organic system that involves various aspects such as energy, environment, society, and economy. Its development requires the support of multiple disciplines and multiple fields. Examples of cooperative demand-side management (DSM) (Razzaq et al., 2016), smart communication infrastructures (Yan et al., 2012), supervised disaggregation model based on deep neural network (Çavdar & Faryad, 2019), and autonomous energy management systems (Gao et al., 2015), are good examples that suggest cooperative design opportunities for better smart grid models. China's smart grid construction should combine its own characteristics, strengthen unified management, and make reasonable top-level designs (Cheshmehzangi, 2020a). The smart grid development should be based on strong grid infrastructure, supported by information and communication platforms, as well as intelligent control and management, to build a strong, reliable, cost-effective, clean, and environmentally friendly modern power grid.

(vii) **Ensuring regulatory and commercial frameworks enable and support the deployment of smart technologies and new commercial practices**—One key challenge of smart grids is that their benefits are undeniable but diffused and challenging to define. Therefore, one critical policy response is to devise a regulatory framework that clarifies these benefits and helps ensure that they flow to the entities providing the upfront investment to incentivise a transition (Agrell et al., 2013; Amin, 2015; Livieratos et al., 2013; Luthra et al., 2014). For example, a utility can be expected to invest in smart grid technologies only if they are confident that the benefits of that investment (notably enabling higher renewable penetration) come with financial benefits (Park et al. 2017). Therefore, regulators must put in place financial incentives structures that appropriately reward smart grid investments. Meanwhile, it also requires an effective price control system and monitoring and communication systems to realise immediate benefits.

(viii) **Motivating consumers to engage in renewable energy transition and smart grid development**—The Government and industry should ensure consumers understand and are convinced by the benefits of renewable energy transition and smart grid development (Cantarero, 2020; Hossain et al., 2016; Worighi et al., 2019). They also need to develop a greater understanding of what motivates consumers to change the way they use energy and which bodies are best placed to engage with consumers. Meanwhile, efforts also should be made to ensure the opportunities to present the benefits of smart grids alongside the rollout of smart meters are maximised.

(ix) **Promoting international collaboration**—As many countries have conducted smart grids development and other smart systems for low carbon development, there is a need for increased coordination between China and other countries to enable communications of technologies and experiences. Greater international collaboration in sharing experiences of pilot programmes (Cheshmehzangi et al., Cheshmehzangi, Xie, et al., 2018; Raven et al., 2019) and in leveraging national investments (Hildebrandt et al., 2017; Lo & Broto, 2019; Matsuo & Schmidt, 2017) in the development of required technology could help optimise and accelerate both the development and deployment of necessary technology. At the same time, such collaborative opportunities could help reduce costs for all stakeholders, namely governments, industry, and the public.

(x) **Improving the safety and security of smart systems**—The development of smart systems (such as smart grid systems) brings additional cybersecurity issues, primarily due to the communications layer's existence, which is responsible for communicating all participants in the electricity network.

Because of the massive dependence of the smart systems on information technology, security problems such as network attacks and information disclosure that occur on the Internet may also appear in the smart system network (Knapp & Langill, 2014; Saxena et al., 2017; Boukerche & Coutinho, 2018; Serror et al., 2018; Cheshmehzangi et al., 2021). As highlighted by Delgado-Gomes et al., (2015, p. 534), "*the main security objectives of the ICT/communications layer are confidentiality, integrity, authorisation, and authentication of the exchanged data*". Network security measures need to consider the equipment and network architecture, various delay constraints, and limited computing resources (Yet et al., 2008). In addition, users' privacy is receiving increasing attention and needs to be further protected to avoid the abuse of information. In the future construction of the smart grid information transmission platform (Wu et al., 2011), attention should be paid to the design of a flat structure, multi-layered distribution, functional grouping, and flexible arrangement. It is necessary to take into consideration the concurrent development of existing communication networks and the parallel development of the emerging network architecture to build a safe, reliable, stable, applicable, and fast smart grid information exchange platform.

11.3 Low Carbon and Climate-Resilient Paradigms

To conclude this chapter, we summarise ten low carbon and climate-resilient urban development paradigms that we believe are important for China's next step of low carbon transitions. We narrate them as key aspects to meeting the next steps of the low carbon transition, which will be challenging for China. In achieving carbon neutrality in the next four decades, many steps should start now, many institutions should be established, and many policies should be reformed. These are just the inception of actions that are meant to create tangible transitions towards a low carbon country. The following paradigms were initially highlighted by Cheshmehzangi et al. (Cheshmehzangi, Li, et al., 2018) and are now further extended as part of the longer-term strategies towards the 2060 carbon neutrality plan.

(i) ***Sustainable low carbon urban transport.*** In short-, medium-, and long-term land use, transport and lifestyle principles of transit-oriented development (TOD) with regional commuter rail, subway, BRT, bus priority transportation should be implemented in intercity and intra-urban public transport. This is suggested to be done along with intelligent transport system technology (ITS). In the short-term, policies may prioritise new-energy vehicles and the expansion of charging facilities and infrastructure. Cities should be encouraged to explore new operation modalities, including customised transportation, shared transportation, and cooperation between the government and the private sector. In the long-term, we have to invest in infrastructural changes and the transformation of urban forms, spatial planning, and planning systems.

(ii) ***Resilient city support and guidelines.*** While the policy of urban adaptation to climate change has been formalised, short-term support to pilot cities is crucial. The experimentation of resilient city development should soon scale up to cover larger areas of neighbourhoods, districts, and cities. In doing so, we can optimise city operations and various networks of the supply chain, construction, logistics, etc. In the medium-term, more support, guidelines, and standards are needed to enable mainstreaming across China.

(iii) ***Sponge City Program (SCP)***—In the short-term, the programme needs to be improved, including classification guidelines and consideration of full catchment areas and long-term risk of climate change in the pilot cities. The next steps require extensive coordination of multiple departments and experts to ensure achieving high-quality urbanisation through sponge city practices (Sun et al., 2020). The main suggestion is for cities to assess their local conditions, research for computability of the programme in their local context, and refine a localised model where needed.

(iv) ***Effective urban disaster risk management*** is urgently needed. Current urban construction in some areas is unable to withstand some potential disaster risks, and greater challenges occur due to climate change impacts. The ongoing pandemic also helps to revisit the risk management systems and the consideration of holistic methods and mechanisms to enhance the city's resilience and responsiveness to disasters and emergencies.

(v) ***Environmental protection planning*** is needed in the short-term and implementation in the medium- and long-term. These should be done to further strengthen the protection of green open space, forests, and wetlands, improving ecosystem services contributing to adaptation to and mitigation of climate change by increasing the capacity of forest carbon sinks and integrating ecosystem services to benefit from water, air, and climate services. The protection of biodiversity and ecosystem services in and outside cities are important aspects for future considerations. More importantly, expansion on the ecological understanding of cities, environmental impact assessments, and environmental protection guidelines are crucial steps for future planning procedures and reforms.

(vi) ***Shared transportation industry*** development is challenging and legal, and policy framework development is slow. Therefore, we see there is a lack of prioritisation of public space for shared vehicles. This requires long-term development, which would reflect on changes in urban form and spatial planning. In this capital-driven market, tailor-made standardisation and guidelines are needed as a framework for management and operation and enable entrepreneurial and user risk management.

(vii) ***Policy support for renewable energy*** and mechanisms to enable decentralised production and supply should be further developed. In doing so, we can consider the short-term development of renewable energy policies and promote wind and solar energy and others. The power generation base is far from the eastern power load center, and it is currently not conducive to the consumption of renewable energy. While China has expanded its renewable energy capacity, the focus should now be on specific mechanisms for production, distribution, and usage.

(viii) ***Smart grid development*** needs support in the short-term, including big data processing technology and knowledge, which need to be improved and integrated with the optimisation of city operations. This needs further financial support and long-term planning for integration in the existing infrastructure and planning system in the mid-to-long term.

(ix) ***Policy agenda inspired by international cooperation plan.*** This is considered in three areas. The first is to provide a range of tools that make cities more resilient. Some examples include local leaders' work manuals, local government self-assessment tools, other tools and experience exchange websites, standardisation, and cities' guidance to be more resilient. These can be developed through experimental pilot cases in the short-term. The second, by organising media activities to provide communication tools and a city resilience analysis report for local governments in a short-to-medium term. And the third is to work with training centers and research institutes to develop suitable risk and robustness training programs for local cities in the medium term. The main actors in charge in the short-term and mid-term to develop these recommendations are the Department of Housing Construction, the Department of Transportation, and other associated departments. In the long-term, they should be developed by all these ministries and the bureau of supervision, and local municipal management bureaus.

(x) ***Regional planning methods and utilising city cluster mechanisms.*** The role of regional planning should be highlighted to "*illustrate the hegemony of urban entrepreneurialism*", and particularly, strategic plans across the nation (Hu, 2016, p. 153). To ensure urban growth and high-quality urbanisation are accomplished, the new-type urbanisation plan (NUP) ideals should be achieved in the medium-to-long term (2020b; Cheshmehzangi, 2014, 2016). The emphasis on regional planning would create a robust ecosystem of urban-rural relations (Cheshmehzangi et al., 2021), which will be essential for better infrastructural systems, spatial networks, green and cleaner production, renewable energy opportunities, supply chains, economic growth, and social development. The approach towards city cluster development was proposed in the 13th FYP (Zhao, 2016), including the proposal for 19 city clusters in China. If further steps are to be taken, the city cluster mechanism should be carefully addressed to reverse the recent and current challenges of severe pollution (Li et al., 2018; Yang et al., 2017), air and water quality issues (Shao et al., 2006), inequalities (Pan et al., 2020), health risks (Wang et al., 2021), and economic productivity (Liu et al., 2018). In a way, we can say the idea of city clusters would only be successful if comprehensive regional planning is done properly to consider multiple aspects of low carbon development, production, system networks, infrastructure development, etc.

11.4 Summary

This chapter highlighted some of the critical aspects, lessons, and paradigms for meeting the next steps of low carbon development. The low carbon transition and climate-resilient progress, central to this book's aims, are crucial in meeting the interim and ultimate steps ahead in the years 2030 and 2060. As we summarised in Chap. 11, China's recent progress is applaudable, but many areas still require further attention, reform, and support (e.g., financially, politically, and institutionally). While we tried to cover a holistic overview of lessons and paradigms for China's next steps in its low carbon transitions, we could only cover areas that we see crucial in what could become the realities of the 2030 targets and the 2060 plan. As we highlighted in this chapter and the previous two chapters, there are still many challenges ahead, and some may become critical in experimentation phases, scaling up procedures, adaptation and mitigation strategies, and restructuring mechanisms. We believe, however, the following steps are essential in order to what could become pathways towards sustainability transitions. In the next chapter, we aim to summarise our key findings and contribute to the ongoing debate, research, policies, and practices of low carbon development. In doing so, we will conclude the book with some key suggestions and steps ahead.

Acknowledgments Ali Cheshmehzangi acknowledges the Asian Development Bank (ADB) and the National Development Reform Commission (NDRC) of China for the opportunity of a research project that he led during 2017 and 2018. In particular, we acknowledge colleagues and friends from ADB for their support and the opportunity to review China's 13th Five-Year-Plan, which then identified the critical areas of low carbon development for the 14th Five-Year-Plan. Besides, we thank our team members at 'Chinese Society for Urban Studies (CSUS)' and 'National Center for Climate Change Strategy and International Cooperation (NCSC)', both based in Beijing, China. In particular, we acknowledge the support of Dr. Hailong LI (leading the CSUS Team) and Dr. Xiu YANG (leading the NCSC Team). With their support and valuable collaboration, we managed to put together a very detailed review and a comprehensive technical report.

References

Agrell, P. J., Bogetoft, P., & Mikkers, M. (2013). Smart-grid investments, regulation and organization. *Energy Policy, 52*, 656–666.

Amin, M. (2015). Smart grid. *Public Utilities Fortnightly*. Available from http://www.ourenergypolicy.org/wp-content/uploads/2015/06/20150604091846-Amin-Maaterials-PUF-1503.pdf.

Andersen, M. S. (1994). *Governance by green taxes: Making pollution prevention pay*. Manchester University Press.

Bai, X., Wieczorek, A. J., Kaneko, S., Lisson, S., & Contreras, A. (2009). Enabling sustainability transitions in Asia: the importance of vertical and horizontal linkages. *Technological Forecasting and Social Change, 76*(2), 255–266.

Baeumler, A., Ijjasz-Vasquez, E., & Mehndiratta, S. (Eds.). (2012). *Sustainable low-carbon city development in China*. World Bank Publications.

Bohnsack, R. (2018). Local niches and firm responses in sustainability transitions: The case of low-emission vehicles in China. *Technovation, 70*, 20–32.

Boukerche, A., & Coutinho, R. W. (2018, June). Smart disaster detection and response system for smart cities. In *2018 IEEE Symposium on Computers and Communications (ISCC)* (pp. 01102–01107). IEEE.

Butters, C., Cheshmehzangi, A., & Sassi, P. (2020). Cities, energy and climate: Seven reasons to question the dense high-rise city. *Journal of Green Building, 15*(3), 197–214.

Cantarero, M. M. V. (2020). Of renewable energy, energy democracy, and sustainable development: A roadmap to accelerate the energy transition in developing countries. *Energy Research & Social Science, 70*, 101716.

Çavdar, İH., & Faryad, V. (2019). New design of a supervised energy disaggregation model based on the deep neural network for a smart grid. *Energies, 12*(7), 1217.

Cheshmehzangi, A. (2014). The urban and urbanism: China's new urbanization and approaches towards comprehensive development. *The International Journal of Interdisciplinary Environmental Studies, 8*(3–4), 1–12.

Cheshmehzangi, A. (2016). China's new-type urbanisation plan (NUP) and the foreseeing challenges for decarbonization of cities: A review. *Energy Procedia, 104*, 146–152.

Cheshmehzangi, A. (2020a). Towards urban energy management: A brief review of china's recent transition in urban energy demand. *Low Carbon Economy, 11*(1), 1–24.

Cheshmehzangi, A. (2020b). Low carbon transition at the township level: feasibility study of environmental pollutants and sustainable energy planning. *International Journal of Sustainable Energy* 1–27. https://doi.org/10.1080/14786451.2020.1860042.

Cheshmehzangi, A., & Butters, C. (Eds.). (2017). *Designing cooler cities: Energy, cooling and urban form: The Asian perspective*. Palgrave Macmillan.

Cheshmehzangi, A., & Dawodu, A. (2019). *Sustainable urban development in the age of climate change—People: The cure or curse*. Palgrave Macmillan.

Cheshmehzangi, A., Dawodu, A., & Sharifi, A. (2021). *Sustainable Urbanism in China*. Routledge.

Cheshmehzangi, A., Xie, L., & Tan-Mullins, M. (2018a). The role of international actors in low-carbon transitions of Shenzhen's international low carbon city in China. *Cities, 74*, 64–74.

Cheshmehzangi, A., Li, H., & Yang, X. (2018b, October). Low Carbon and Climate Resilient urban development in China. ADB Report, developed for the National Development Reform Committee (NDRC), Beijing, China.

Delgado-Gomes, V., Martins, J. F., Lima, C., & Borza, P. N. (2015, June). Smart grid security issues. In *2015 9th International Conference on Compatibility and Power Electronics (CPE)* (pp. 534–538). IEEE.

Gao, B., Liu, X., Zhang, W., & Tang, Y. (2015). Autonomous household energy management based on a double cooperative game approach in the smart grid. *Energies, 8*(7), 7326–7343.

Hajibabai, L., & Ouyang, Y. (2013). Integrated planning of supply chain networks and multi-modal transportation infrastructure expansion: Model development and application to the biofuel industry. *Computer-Aided Civil and Infrastructure Engineering, 28*(4), 247–259.

Harris, I., Wang, Y., & Wang, H. (2015). ICT in multimodal transport and technological trends: Unleashing potential for the future. *International Journal of Production Economics, 159*, 88–103.

He, B. J., Zhao, D. X., Zhu, J., Darko, A., & Gou, Z. H. (2018). Promoting and implementing urban sustainability in China: An integration of sustainable initiatives at different urban scales. *Habitat International, 82*, 83–93.

Hildebrandt, J., Hagemann, N., & Thrän, D. (2017). The contribution of wood-based construction materials for leveraging a low carbon building sector in Europe. *Sustainable Cities and Society, 34*, 405–418.

Hossain, M. S., Madlool, N. A., Rahim, N. A., Selvaraj, J., Pandey, A. K., & Khan, A. F. (2016). Role of smart grid in renewable energy: An overview. *Renewable and Sustainable Energy Reviews, 60*, 1168–1184.

Hu, R. (2016). Planning for growth: Urban and regional planning in China. *Planning Perspectives, 31*(1), 152–155.

Huang, P. (2019). The verticality of policy mixes for sustainability transitions: A case study of solar water heating in China. *Research Policy, 48*(10), 103758.

Knapp, E. D., & Langill, J. T. (2014). *Industrial Network Security: Securing critical infrastructure networks for smart grid, SCADA, and other Industrial Control Systems*. Syngress.

Li, X., Wang, Y., Shen, L., Zhang, H., Zhao, H., Zhang, Y., & Ma, Y. (2018). Characteristics of boundary layer structure during a persistent haze event in the central Liaoning city cluster, Northeast China. *Journal of Meteorological Research, 32*(2), 302–312.

Lin, Z., & Yang, H. (2002). Impact analysis of advanced traffic information system on travel choice. *Communication and TransportatiOn Systems Engineering and Information* (Vol. 4). (In Chinese).

Livieratos, S., Vogiatzaki, V. E., & Cottis, P. G. (2013). A generic framework for the evaluation of the benefits expected from the smart grid. *Energies, 6*(2), 988–1008.

Lo, K., & Broto, V. C. (2019). Co-benefits, contradictions, and multi-level governance of low-carbon experimentation: Leveraging solar energy for sustainable development in China. *Global Environmental Change, 59*, 101993.

Luthra, S., Kumar, S., Kharb, R., Ansari, M. F., & Shimmi, S. L. (2014). Adoption of smart grid technologies: An analysis of interactions among barriers. *Renewable and Sustainable Energy Reviews, 33*, 554–565.

Matsuo, T., & Schmidt, T. S. (2017). Hybridizing low-carbon technology deployment policy and fossil fuel subsidy reform: A climate finance perspective. *Environmental Research Letters, 12*(1), 014002.

Oates, W. E. (1995). Green taxes: Can we protect the environment and improve the tax system at the same time? *Southern Economic Journal*, 915–922.

Pan, W., Fu, H., & Zheng, P. (2020). Regional poverty and inequality in the xiamen-zhangzhou-quanzhou city cluster in China based on NPP/VIIRS night-time light imagery. *Sustainability, 12*(6), 2547.

Park, C., Kim, H., & Yong, T. (2017). Dynamic characteristics of smart grid technology acceptance. *Energy Procedia, 128*, 187–193.

Raven, R., Sengers, F., Spaeth, P., Xie, L., Cheshmehzangi, A., & de Jong, M. (2019). Urban experimentation and institutional arrangements. *European Planning Studies, 27*(2), 258–281.

Razzaq, S., Zafar, R., Khan, N. A., Butt, A. R., & Mahmood, A. (2016). A novel prosumer-based energy sharing and management (PESM) approach for cooperative demand side management (DSM) in smart grid. *Applied Sciences, 6*(10), 275.

Saxena, N., Grijalva, S., Chukwuka, V., & Vasilakos, A. V. (2017). Network security and privacy challenges in smart vehicle-to-grid. *IEEE Wireless Communications, 24*(4), 88–98.

Serror, M., Henze, M., Hack, S., Schuba, M., & Wehrle, K. (2018, August). Towards in-network security for smart homes. In *Proceedings of the 13th International Conference on Availability, Reliability and Security* (pp. 1–8).

Shao, M., Tang, X., Zhang, Y., & Li, W. (2006). City clusters in China: Air and surface water pollution. *Frontiers in Ecology and the Environment, 4*(7), 353–361.

Wang, C., Luo, H., Zhao, F., & Qin, Y. (2020). Combining residual and LSTM Recurrent networks for transportation mode detection using multimodal sensors integrated in smartphones. *IEEE Transactions on Intelligent Transportation Systems.*

Wang, M., Lyckvi, S. L., & Chen, F. (2016, October). Same, same but different: how design requirements for an auditory advisory traffic information system differ between Sweden and China. In *Proceedings of the 8th International Conference on Automotive User Interfaces and Interactive Vehicular Applications* (pp. 75–82).

Wang, Y., Fang, X., Yin, S., & Chen, W. (2021). Low-carbon development quality of cities in China: Evaluation and obstacle analysis. *Sustainable Cities and Society, 64*, 102553.

Worighi, I., Maach, A., Hafid, A., Hegazy, O., & Van Mierlo, J. (2019). Integrating renewable energy in smart grid system: Architecture, virtualization and analysis. *Sustainable Energy, Grids and Networks, 18*, 100226.

Wu, Y. N., Chen, J., & Liu, L. R. (2011). Construction of China's smart grid information system analysis. *Renewable and Sustainable Energy Reviews, 15*(9), 4236–4241.

Xie, H. (2014). Legal regulation of low-carbon economy. *IERI Procedia, 8*, 170–175.

Xing, Z. (2018). Low-carbon cities evaluation model based on RS and SVM. *Journal of Applied Science and Engineering Innovation, 5*(2), 51–54.

Yang, T., Gbaguidi, A., Yan, P., Zhang, W., Zhu, L., Yao, X., Wang, Z., & Chen, H. (2017). Model elucidating the sources and formation mechanisms of severe haze pollution over Northeast mega-city cluster in China. *Environmental Pollution, 230*, 692–700.

Yang, X., Wang, X. C., & Zhou, Z. Y. (2018). Development path of Chinese low-carbon cities based on index evaluation. *Advances in Climate Change Research, 9*(2), 144–153.

Yap, X. S., & Truffer, B. (2019). Shaping selection environments for industrial catch-up and sustainability transitions: A systemic perspective on endogenizing windows of opportunity. *Research Policy, 48*(4), 1030–1047.

Zhao, L. (2016). China's 13th Five-Year Plan: Road Map for Social Development. *East Asian Policy, 8*(03), 19–32.

Zhao, Z. Y., Gao, L., & Zuo, J. (2019). How national policies facilitate low carbon city development: A China study. *Journal of Cleaner Production, 234*, 743–754.

Chapter 12
Key Suggestions and Steps Ahead for China's Carbon Neutrality Plan

12.1 Introduction

As a major carbon emitter and coal power producer, China's net-zero carbon commitment will reshape its economy and contribute to global decarbonisation and climate action. China's ambition to be carbon neutral by 2060 will reduce global temperatures by about 0.2 to 0.3 °C below forecast (Climate Action Tracker, 2020). This chapter serves as the concluding chapter of the book. However, instead of providing some generic concluding remarks, we aim to highlight key suggestions and steps ahead for China's carbon neutrality plan. In the face of uncertainties after COVID-19, China needs to come up with a more resilient, low carbon, and sustainable long-term economic development plan in the 14th Five-Year Plan period (2021–2025), including a renewed Nationally Determined Contribution (NDC) and a Long-term Low Greenhouse Gas Emission Development Strategy (LTS) (EF, 2020; WRI, 2020). The 2060 Carbon Neutral Commitment remains an overarching principle rather than a detailed road map. The details need to evolve as economic needs, the international environment, and technological developments change. We believe we can touch on key aspects that should be considered in the next steps of reaching the 2030 targets and the 2060 goals (Fig. 12.1).

This chapter serves as the concluding chapter of the book. However, rather than a generic conclusion, we delve into some practical suggestions for the steps ahead. Therefore, we first explore China's international position, as we believe it will become more important in the following years. Despite the negative impacts of the COVID-19 pandemic on China's trade and economic relations (Cheshmehzangi, 2020a; Cheshmehzangi, 2021), we see a growing demand for infrastructural development, large-scale urban projects, and further cooperation and investment between China and other countries. Here, we explicitly look into the Belt and Road Initiative (BRI) as a major platform for international cooperation, overseas investment, and regional cooperation development (Teo et al., 2019). Afterward, we delve into further discussions on China's sustainability transitions, echoing the important factors related to the book's main aims and objectives. We then provide two sections on China's deep

A. Cheshmehzangi and H. Chen, *China's Sustainability Transitions*,
https://doi.org/10.1007/978-981-16-2621-0_12

Fig. 12.1 The path to China's carbon-neutral vision (Drawn by Authors)

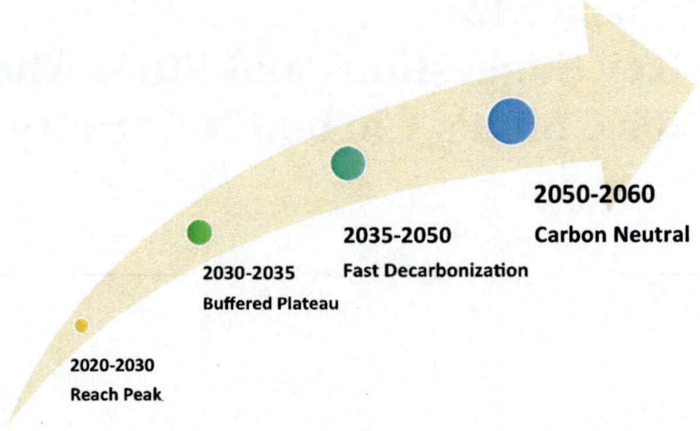

decarbonisation plan from 2021–2030 and its path towards the 2060 carbon neutrality plan. In doing so, we then summarise ten suggestions for the next steps of China's sustainability transitions. In this closing section, we explore key areas that require long-term vision, investment, and planning. This serves as the book's concluding remark, highlighting the importance of low carbon and climate-resilient directions towards the ultimate goals of 2060.

12.2 The Belt and Road Initiative: Embark on Green Co-development

In recent years, China has become a major international player that suggests new pathways towards cooperation, collaboration, investment, trade, and development opportunities. Since 2017, a major shift has taken place to expand green development opportunities and international engagement in such projects. One of the major shifts was related to the earlier global initiative, namely the Belt and Road Initiative (BRI). This initiative has become an international norm in promoting green development strategies in specific regions (Rio & Firza, 2021). Since 2017, the BRI has specific green objectives (Zhou et al., 2018a, 2018b), showing opportunities for competitiveness, convergence, and regional development (Zhang et al., 2020). Many arguments now exist around the role of BRI and its effectiveness in achieving global low carbon development (Chen et al., 2020) or to achieve sustainable development in specific regions (Liu, 2018). Some studies identify the shift of BRI towards developing a new model for green and sustainable investment (Yin, 2019). If so, we believe China has done well in establishing and rebranding the BRI as a major platform on green co-development. In this regard, the BRI could be identified as a sustainable model of development, through which the targets of sustainable development goals (SDGs), green development, and cleaner productions could not be neglected.

Since 2013, the Chinese government has issued a series of policy documents to strengthen the BRI ecological and environmental protection, such as the Guidelines on Environmental Protection for Overseas Investment and Cooperation (Ministry of Commerce, 2013). In 2015, the government published the "Vision and Actions on Jointly Building Silk Road Economic Belt and 21st-Century Maritime Silk Road", requiring enterprises to take the initiative to shoulder social responsibilities and strictly protect biodiversity and the ecological environment (Ministry of Commerce, 2015). In 2017, the government issued some important documents such as "Guidance on Promoting Green Belt and Road" (MEE et al., 2017) and "Belt and Road Ecological and Environmental Cooperation Plan" (MEE, 2017), which have clarified the overall route and tasks of green BRI. China then issued the "Action Plan on Belt and Road Standard Connectivity (2018–2020)" which has signed 85 standardisation cooperation agreements with 49 countries and regions. Chinese and British institutions jointly released the Belt and Road Initiative Principles for Green Investment in 2018. It has received 37 signatories and 12 supporting institutions from 14 countries and regions (Belt & Road Portal, 2018). The International Coalition for Green Development on the Belt and Road (XinhuaNet, 2019), which is established in 2019, has created a cooperative communication platform for Green Development.

12.3 China's Sustainability Transitions: Low Carbon and Climate-Resilient Directions

The rise of carbon neutrality may appear as a carbonwash/greenwash idea (Rauland & Newman, 2015), but in reality, it provides longer-term plans as part of sustainability transitions. It helps formulate steps towards the development of new guidelines, frameworks, and protocols, especially for standardisation, certification, regulations, etc. While no universal definition exists, we urge to understand how carbon neutrality should happen through low carbon transition(s) and climate-resilient directions. By planning carbon–neutral built environments, we move closer to work with relevant stakeholders and industries (Cheshmehzangi et al., 2018; Zuo et al., 2012) and create coherent development patterns or trends that could lead to low carbon planning and design, as well as low carbon standards, methodologies, and technologies. In examining what could be regarded as a carbon–neutral model, we define a wide range of opportunities and challenges. These could be especially related to sustainable urban development frameworks (Kennedy & Sgouridis), climate change mitigation strategies (Pasimeni et al., 2019), life cycle perspectives (Hendrickson et al., 2016), and incorporating various factors that help us develop a sustainable urban systems framework (Jo et al., 2009). Popular methods of GHG emission abatement strategies (Dai et al., 2020; Ju et al., 2021) are emerging fast. More recent examples include the integration of renewable and low carbon energy technologies (Janzen et al., 2020), air pollution decontamination (Paardekooper et al., 2020), and shifting towards global climate governance (Obergassel e al., 2021), which are just a few emerging examples of effective decarbonisation and low carbon strategies. Indeed, such methods

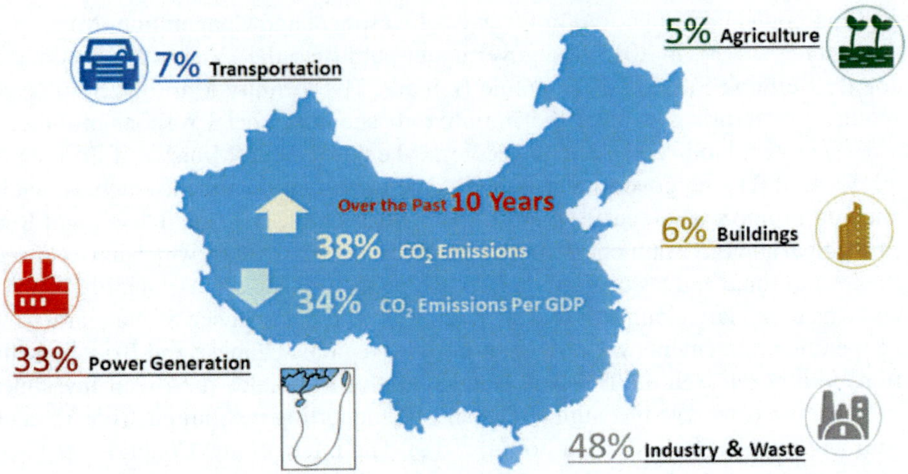

Fig. 12.2 China CO_2 emission and GHG emission split (Drawn by the Authors, data is extracted from Goldman Sachs Research, 2021)

and strategies require long-term visioning and planning, and therefore, a systematic progression is ever needed (Fig. 12.2).

12.4 China's 2021–2030 Plan Period: Deep Decarbonisation

First of all, in the critical decade of the 2030 carbon peak target, China needs to peak its carbon emissions as early as possible and strictly control the peak of its emissions. In doing so, China could leave more time for the reduction process from high-quality peak to carbon neutral, make policy formulation, energy structure reform and deployment, and transform production and lifestyle more conveniently (Wang & Zhang, 2020). Policymakers and stakeholders such as producers and consumers need to work together on these three major tasks by 2030. One is to accelerate the deployment of zero-carbon solutions and build the infrastructure and networks to do so. Second, it is necessary to create a reasonable policy and investment environment to promote the dissemination and introduction of new technology. The third is to provide research and development support and risk mitigation to ensure the emergence of zero-carbon technologies in the future (ETC, 2020).

Over the next five years, efforts will focus on sectors that are driving the transformation: renewable energy, electric vehicles, batteries and storage, and hydrogen (The Oxford Institute for Energy Studies, 2020). The National Energy Technology Model (NET) developed by the Beijing Institute of Technology confirms that the annual growth rate of China's carbon dioxide emissions should be no more than 0.4% from 2020 to 2030, and the annual increase of carbon dioxide emissions should be controlled within 50 million tons to achieve the carbon peak target. Among them, the direct CO_2 emission reduction responsibility of energy system industries is 49% of electric power and thermal power, 29% of the industry, 12% of transportation, and 10% of building (Yu et al., 2021). According to Energy Policy Solutions (EPS), China has the potential to achieve more ambitious NDC targets, including an early peak in CO_2 emissions, an increase in the share of non-fossil energy in primary energy consumption, and a higher rate of reduction in carbon intensity. EPS results also show that during the 14th Five-Year Plan period, China's carbon dioxide emissions should be controlled at about 10.3 billion tons, carbon dioxide emissions per unit of GDP should be reduced by about 22% compared to 2020, and non-fossil energy should account for about 20% of primary energy consumption (WRI, 2020). Advanced regions such as the Beijing-Tianjin-Hebei region, the Yangtze River Delta region, and Guangdong, Hong Kong, and Macao could lead China's low carbon transition if they reach carbon dioxide peak early during the 14th Five-Year Plan period.

Moderate Energy Revolution and Transformation

China has seen the largest increase in energy efficiency and the fastest decline in energy intensity globally, with an average annual decline rate of more than 2 percent (Zou et al., 2021). China is also the country with the fastest development and largest utilisation of new energy in the world. However, the energy transformation needs to be measured. Lin and Zhu (2021) put forward the assumption that the current total carbon dioxide emission from energy consumption in China is calculated at 10 billion tons. If total energy consumption in 2030 is 6 billion tons of standard coal, as shown in Fig. 12.3, non-fossil energy consumption will account for 25%. That would mean 10.5 billion tons of carbon dioxide emissions from energy consumption in 2030, which would make it possible to reach a carbon peak before 2030. In 2035,

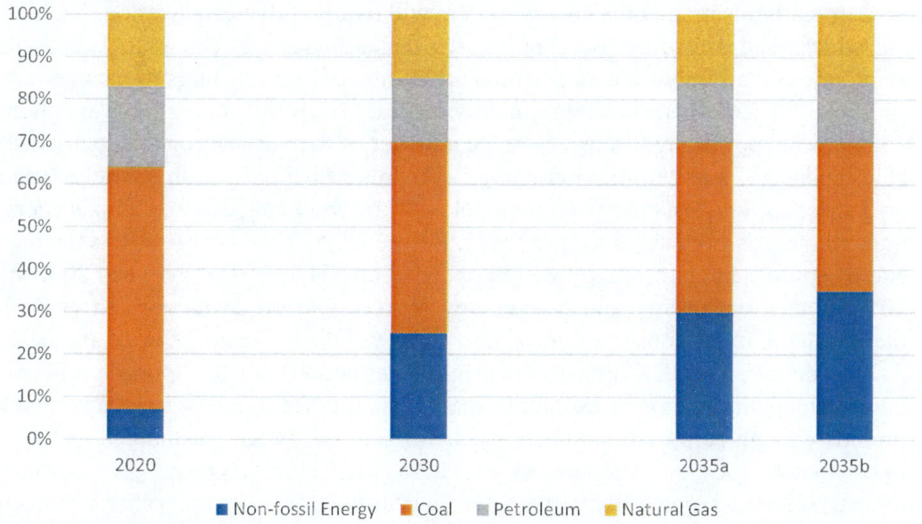

Fig. 12.3 The projected share of energy consumption at carbon peaks (Drawn by the Authors)

the proportion of non-fossil energy consumption will be 30%-35% (at this time, the carbon dioxide emissions will be between 8.9 and 9.7 billion tons), which can ensure that the carbon dioxide emissions from energy consumption will drop significantly after the carbon peak in 2030, and gradually enter carbon neutrality.

12.5 Path to 2060 Carbon Neutrality Goal

The report by the Institute of Climate Change and Sustainable Development at Tsinghua University showed that China could achieve its carbon–neutral target by 2060 or earlier, on track to keep global temperature rise to 1.5 °C. In the process, China will reach net-zero carbon dioxide emissions by 2050 and reduce all greenhouse gases by 90% (ICCSD, 2020). According to Goldman Sachs Research (2021), it is estimated that by 2060, China's net-zero emissions will lead to $16 trillion in clean technology infrastructure investment opportunities and about 40 million net new jobs. At the same time, China's electricity generation will triple, driven mainly by solar, wind, nuclear and hydro.

Similar to the model of carbon neutrality process proposed by Boake (2008), we emphasise the need for a continuous cycle of measuring, reducing, renewing, and offsetting is important in achieving the proposed carbon neutral plan. While sector-specific frameworks are encouraged (Rauland & Newman, 2015), we also suggest considering multi-sectoral and multi-objective development that is not just limited to carbon neural concepts and initiatives in individual sectors. The emission path under the carbon–neutral vision relies on the technical support of three major fields: energy-efficient recycling, zero-carbon energy, and disruptive negative emissions (Wang & Zhang, 2020) (see Fig. 12.4 for more details).

Zero-carbonisation of the energy system will require full electrification as a basis, the widespread use of zero-carbon energy technologies and processes across the economy, and an important shift from carbon-intensive fossil fuels to clean energy (ETC, 2020). Zero-carbon energy mainly includes renewable energy power (photovoltaic, wind power, hydropower), nuclear power, zero hydrocarbon energy, sustainable bioenergy, and zero-carbon energy comprehensive utilisation services (smart grid, electric vehicles, energy storage), etc. (Wang & Zhang, 2020). China's energy transition needs to be achieved gradually. Before 2035, fossil energy still is the main energy source, and new energy will be speeded up (ETC, 2021). Between 2035 and 2050, both fossil energy and new energy sources will be developed. After 2050, the production and consumption of new energy will dominate (Zou et al., 2021). At the technical level, China's industrial, construction and transportation sectors have many path options for eventual decarbonisation. But there is an urgent need to provide enough zero-carbon electricity, hydrogen energy and sustainable biomass, or use matching Carbon Capture use and Storage (CCUS). At present, in the cement industry to reduce power loss, synthetic fuels used in aviation and other aspects of the technology is not mature (ETC, 2019). Carbon capture plays an important role in heavy industry, but the cost in power systems is still high.

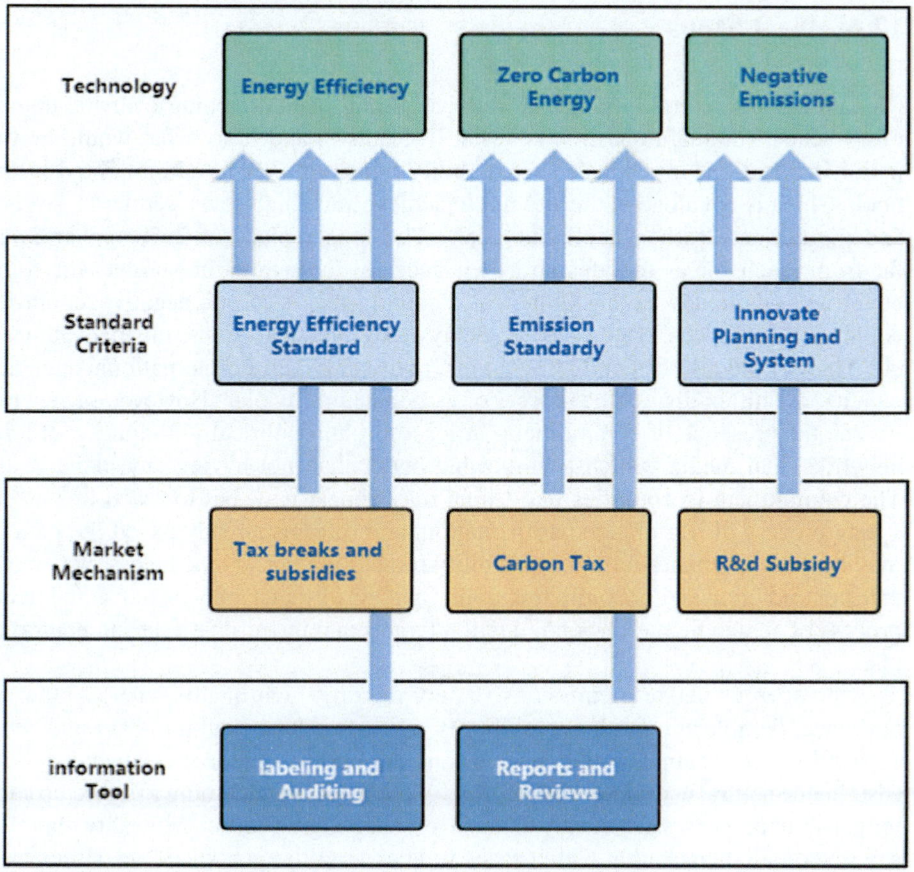

Fig. 12.4 A policy system based on carbon neutral (Drawn by the Authors, using the translated data on factors and highlights of the study by Wang & Zhang, 2020)

During this period, China should phase out all coal-fired power stations that do not meet effective carbon reduction targets. In addition, the economic dispatching across the regional power market will promote the reform of the power market in the future. China also needs to implement low carbon strategies in the steel, chemical and cement sectors, and provide routes to electricity, hydrogen and carbon capture and storage (CCS) production technologies. Eliminate the sale of fossil fuel cars and vans as much as possible by 2040 while developing and implementing a comprehensive decarbonisation strategy for transport vehicles. Besides, implementing energy efficiency retrofits for residential, commercial, and public buildings by 2050 for existing buildings (PRI, 2021). Therefore, we can argue in favour of potential opportunities in specific sectors, cross-sectoral approaches, and multi-sectoral development. In the following section, we summarise a set of suggestions elaborating on the next steps of moving towards the ultimate carbon neutrality plan of 2060.

12.6 Next Steps

China's carbon neutrality plan is a relatively long plan stretching over the course of 40 years. The main question is 'what if it fails?'; and if so, what would be the major implications on China's sustainability transitions? As we highlighted in the book, China is not alone in the attempt to achieve such high-level standards, targets, and the ultimate carbon neutrality goals. The longer plans of 2050 onwards are far from reach for us to safeguard their success. Despite earlier targets for some countries and already having Bhutan as a country that is carbon negative, countries could experience challenges such as delays, failures, distortions, mismanagement, etc. There could also be conflicting points, political issues, or international agendas, which may shift or disrupt the process of carbon neutrality plan. However, in general, we see the presence of commitment, at least through political willingness, official networks, and media, which is somewhat better than not having any plans at all. The commitment in countries like China may appear new, but to reach this stage, a long process of learning, experimentation, and reforms already paved the path of moving towards longer-term sustainability transition plans. Nonetheless, we anticipate contradicting and/or conflicting points and views that cannot be neglected in the process of achieving the carbon reduction targets and the ultimate carbon neutrality plan.

A prosperous net-zero-emissions society requires high-quality, energy-efficient buildings, flexible zero-carbon mobility (Carbon-free long-haul transport and zero-carbon logistics chains), zero-emissions circular goods, abundant clean energy, and sustainable natural ecosystems (ETC, 2020). It is always welcoming to speed up such longer-term processes, especially to achieve a challenging carbon neutrality plan. We see several challenges that China faces in achieving the goals, and we summarise some of the overarching ones here. In making a list, we aim to consider primary ideas and pathways that enable China (and other countries) to reach the carbon neutrality goals. The plan, however, is more complex than just a set of suggestions for the next steps. In this regard, we reflect mainly on what we see essential in the next steps of achieving low carbon and climate-resilient targets and toward a carbon neutrality plan. The following ten points are considered vital to the next steps (Fig. 12.5).

12.6.1 Development of Behaviour Change Programmes

China's continuing urbanisation and the aim to go beyond the 75% urbanisation rate by 2030 indicate major challenges in achieving low carbon targets, net-zero goals, and the ultimate carbon neutrality plan. So far, we see China's growing urbanisation is tangled with further industrialisation, rising production, and increasing consumption behaviours. Therefore, it will be essential to consider the development of behaviour change programmes, such as those already initiated for waste management, food waste, recycling, public transportation use, cycling, etc. Such programmes would

Fig. 12.5 Net-zero-emissions steps

require trial evaluation (Hardeman et al., 2005), testing, maintenance, and continuous improvement.

12.6.2 *Gradual But Strategic Reduction of Energy Consumption and Demand*

Reaching the peak in energy consumption and CO_2 emissions may differ across sectors, such as the industrial sector (Zhou et al., 2018a, 2018b) and higher-polluting industries. While there are global and sub-regional strategic considerations for energy supply and demand interdependencies (Matsuo et al., 2013), we see a greater demand for regulatory policies and future evolutionary pathways (He et al., 2016) in achieving a better sustainable development strategy. From the energy situation (Zhang et al., 2011), we require pathways that suggest energy-saving strategies, energy demand reduction, industrial restructuring, and achieving energy efficiency through optimised low carbon models and strategic energy use CO_2 reduction mechanisms (Cheshmehzangi, 2020b). Besides, there is a need for a coordinated system of sub-sector energy transition quantification indicators, such as the share of renewable energy and nuclear power generation (ETC, 2019). Therefore, a gradual but strategic approach is highly recommended to ensure the value of standards, certifications, regulations, incentives, and potential benchmarking systems.

12.6.3 Institutional Rearrangements and Regulatory Enhancement

Similar to the other examples of law on climate change and energy transitions, the next steps also require social and institutional rearrangements (Sorman et al., 2020). Unquestionably, such rearrangements should be energetically backed up by game-changing mechanisms with robust policy and strategies (Tyfield et al., 2010) that enable the rearrangements to be accurate, adequate, and effective. The overarching movements of low carbon economies (García-García et al., 2020) and circular economy (Yong, 2007) are suitable examples of structural rearrangements and strategic solutions towards regulatory enhancement and institutional upgrading opportunities. The former is embedded as part of China's strategic plans in the last decade or so, and the latter is considered a shift from rhetoric plan to practical implementation (Su et al., 2013). In achieving these, we recommend the importance of multiple stakeholders and actors, as well as the careful consideration of developing robust frameworks, systematic diversification and involving new practices.

12.6.4 Investment in New Technologies and Integrating Them for Decarbonisation Plans

Despite earlier uncertainties in low carbon technologies for decarbonisation targets (Kannan, 2009), we see new technologies have recently become important parts of accelerating decarbonisation processes (Kern & Rogge, 2016), innovation (Chen et al., 2009), and towards low carbon futures (Sansom, 2014). In many cases, technological innovation system development is considered effective as part of governing energy transitions, enhancing capabilities and performances (Yam et al., 2011), and suggest the development, diffusion, and use of new technologies (Edsand, 2019). The construction of such a system is not an easy task and requires extensive financial support, especially through the provision of relevant and effective incentives and investment opportunities. In doing so, we suggest the shift towards larger-scale integration of new low carbon technologies in planning and design stages and optimising the built environment performances, enhancing the efficiency of operations, and reducing energy and CO_2 emissions. In terms of energy, China is rich in coal, short of oil and gas, and unstable in the new energy. Therefore, the upgrading of the power generation industry is a reliable guarantee to achieve carbon neutrality. It is significant to promote the innovation of coal-fired power generation (Huang et al., 2021), continuously strengthen the deep peak adjustment ability of coal-fired power generating units (Table 12.1).

Table 12.1 Green transformation technologies for energy systems in various industries (Drawn by the Authors, data is extracted from the study by Yu et al., 2021)

Electric power	Wind power; photoelectricity, Carbon Capture and Storage (CCS)
Iron and steel	Short-term: ball sintering, low temperature sintering, dry quenching coke, dry blast furnace top residual pressure residual heat power generation Mid-long term: large arc furnace steelmaking, hydrogen energy steelmaking and CCS
Chemical	Light raw materials, advanced coal gasification technology, low carbon hydrogen production and CO_2 utilisation technology, CCS
Building	Improve the efficiency of heating and cooling and electrification, and develop distributed energy sources
Traffic	Electric car, hydrogen fuel vehicles, biofuel aircraft and ships

12.6.5 Fast-Forward Experimental Projects for Scale-Up Scenarios and Larger-Scale Development

The initial phase of China's NUP (2014–2020) has not been successful. The suggested reforms are yet to be developed or scaled-up towards larger sustainable development examples. Fast-forwarding experimental projects would enable us to uncover new low carbon development opportunities, such as industrial symbiosis projects (Dong et al., 2014), smart grid systems for low carbon electricity economy (Verbong et al.,), reconfiguration of urban infrastructure (Bulkeley et al., 2010), and produce renewable energies and low carbon fuels. In all cases, we urge to avoid prolonged experimentation but consider longer-term planning and scaled-up scenarios. The impacts would then become more apparent in achieving decarbonisation processes and successful examples of low carbon transitioning and comprehensive low carbon governance. These are likely to be intact with processes of financialisation and dispossession (Sovacool et al., 2019), which should be carefully considered in scaling-up processes.

12.6.6 Considering Revolutionary/Innovative Planning and Design Paradigm Shifts

Although many experimental sustainable projects such as eco-cities and eco-districts, low carbon cities and towns, smart districts and smart grid systems, green communities and green development, etc., the general planning and design practices are mostly unchanged. As suggested by Butters et al. (2020), we need to revisit some of the ongoing development trends to ensure paradigm shifts would become purposeful and effective in the long run. Currently, there is no sign of major paradigm shifts, or there is only a little indication that some of the current planning and design trends are expected to change direction in the near future. We see great potential in areas of resource management, environmental protection, spatial planning, morphological urban design, transportation systems, and industrial restructuring processes. The examples of green or eco-industrial parks, mixed-use development projects, multi-objective planning projects, smart-resilient urban systems, solar cities, and certified green development projects (Cheshmehzangi et al., 2021) are suitable examples of innovative planning and design paradigm shifts. In all cases, we see a sound level of indicators that suggest pathways towards low carbon or net-zero futures.

12.6.7 Build Electricity Market and Carbon Trading Market

It is necessary to construct a mid-to-long-term power market, spot market, auxiliary service market, and capacity market. To achieve a clean and low carbon "premium" for renewable energy through the quota system, green card trading mechanism, and carbon emission right trading mechanism (Lin & Zhu, 2021). China needs to build a unified trading market based on the pilot experience of carbon trading. In addition, it is important to make systematic arrangements in carbon emission quotas, enterprise participation scope, and product pricing mechanism to optimise resource allocation, manage climate risks, and reduce carbon emissions at low cost and high efficiency (Liu, 2021). China can use the national carbon emission trading market to explore a reasonable and more efficient price range for the carbon market and gradually bring industries other than electricity into the carbon market. Besides, the electricity market needs to be promoted in a coordinated way, and the energy use rights trading market and the carbon trading market should be deeply integrated (WRI, 2020). In doing so, China could force the optimisation of the energy structure and explore the space for emission reduction.

12.6.8 Effective Fiscal and Financial Policy Guarantees

Apart from Clean Development Mechanism (CDM) projects and renewable energy tariff surcharges, China's public funding revenues are far from sufficient to support the large sums needed for carbon neutrality. Therefore, it is necessary to continuously improve the investment and financing system and mechanism related to carbon emission reduction in the future and increase the funding sources and financial input to local governments. By setting up a special fund to help and support traditional energy industries or avoid the inequality and other problems resulting from the low carbon transition (Liu, 2021). This mechanism could also help to reduce income inequality (Cevk & Correa-Caro, 2015), enabling the opportunity towards inclusiveness, poverty reductions, and potential decentralisation opportunities. As an example, fiscal decentralisation (Shen et al., 2012) could improve central-local governmental mechanisms. Chin multi-level structure of government, being central → provincial → prefectural → county level → township level, could create opportunities to consider sub-national agenda, create regional targets and assignments, respond to vertical and horizontal fiscal inequalities, and develop tailor-made policies, standards, and benchmarks. These could potentially create better financial support, incentive mechanisms, and collaborative platforms.

12.6.9 Participate in International Carbon Reduction Cooperation

China should play a role in international organisations to achieve a 100% emission reduction target. The International Maritime Organization and the International Civil Aviation Organization, for example, have both set an indicative goal of reducing carbon dioxide emissions by 50% by 2050 (ETC, 2019). More recent studies suggest pathways to international cooperation on climate governance (Liu & Lo, 2020), highlighting, in particular, China's role and collaborating wither other countries, firms, and specific supply chains (Zhang & Wang, 2014). We also see China's response to the Paris Agreement has been relatively positive (Gao, 2016), suggesting the new or upgraded take on green development and low carbon agenda. We see opportunities for China's participation in implementing and advocating green development through existing and potential partnership mechanisms. As mentioned earlier, the BRI is a major platform for China to play its parts in the global arena. Nevertheless, we need to see a shift in enterprises' behaviour, production level, market management, and cooperative climate policy. The latter, although developed after the Paris Agreement, is still yet to be matured. Therefore, we see China's position to become more important than before. International cooperation should become a two-way system of learning and knowledge sharing, and including also potential sub-regional attempts to reduce climate change effects, or country-to-country engagement in achieving low carbon and green development.

12.6.10 Innovation in Key Sectors to Decarbonise the Economy

Innovation financing relies on technologies related to multiple sectors, including electrolytic hydrogen generation, synthetic hydrocarbon production, carbon capture, and storage/utilisation, and biofuels from the most sustainable feedstock sources. The initial phase of commercial-scale deployment of these technologies will require de-risk mechanisms to reduce costs and risks at all steps in the value chain (ETC, 2020) (see Fig. 12.6). There are already reaping economic benefits of decarbonisation (Teng & Joyzo, 2014) that China has managed to consider for its energy system (Burandt et al., 2019), transportation sector (Pan et al., 2018), power sector (Wang et al., 2019), and renewable resources (Khanna et al., 2019). The more recent suggestions

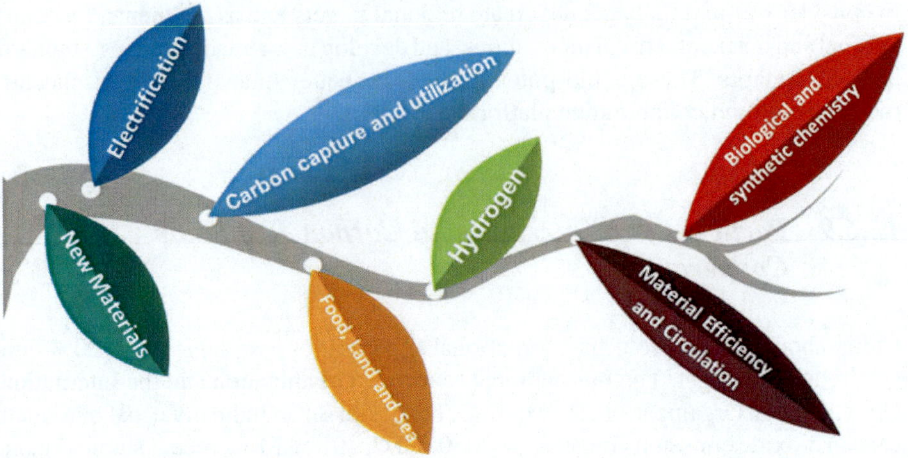

Fig. 12.6 Key Innovation areas of decarbonisation

are to move towards creating a multi-sectoral system or a sort of complex network approach (Ma et al., 2019), which could be effective in pushing for innovation in key sectors. With support from effective driving forces and analysing the economic benefits of decarbonisation, we see more opportunities arising in new technologies, digitalisation, ICTs, blockchain networks, and policy instruments.

Lastly, we value the importance of comprehensive thinking in achieving carbon neutrality goals. As noted in the book, low carbon development includes many businesses, industries, and jobs. We cannot anticipate all of them to do their jobs rightly. Therefore, some companies or industries may take advantage of certain policies, manipulate their data and processes, or even neglect the regulations and guidelines. In this regard, it is inevitable that a percentage of those stakeholders may not be truthful or responsible for their carbon targets, data assessment, and policy procedures. We should also anticipate potential conflicts between different stakeholders, especially in terms of considering the benefits or co-benefits of low carbon measures. In this regard, we note the remaining challenges could become hiccups of the next steps, and we hope relevant stakeholders could overcome them in the current decade. We conclude our points by reflecting on the importance of low carbon transitions, achieving the major goals of sustainability and healthy and harmonious living environments. We recognise the next steps, particularly until 2030, will be very critical to fast-forward a healthy and sustainable transition that could reach the carbon neutrality plan of 2060.

References

Belt, Road Portal. (2018). *Chinese and British agencies jointly release the The Belt and Road initiative principles for green investment*. Available from https://www.yidaiyilu.gov.cn/xwzx/gnxw/73259.htm

Boake, T. M. (2008). The leap to zero carbon and zero emissions: Understanding how to go beyond existing sustainable design protocols. *Journal of Green Building, 6*(4), 64–77.

Bulkeley, H., Broto, V. C., Hodson, M., & Marvin, S. (Eds.). (2010). *Cities and low carbon transitions (Vol. 35)*. Routledge.

Burandt, T., Xiong, B., Löffler, K., & Oei, P. Y. (2019). Decarbonizing China's energy system—Modeling the transformation of the electricity, transportation, heat, and industrial sectors. *Applied Energy, 255*, 113820.

Butters, C., Cheshmehzangi, A., & Sassi, P. (2020). Cities, energy and climate: Seven reasons to question the dense high-rise city. *Journal of Green Building, 15*(3), 197–214.

Cevik, M. S., & Correa-Caro, C. (2015). *Growing (un) equal: fiscal policy and income inequality in China and BRIC+*. International Monetary Fund. IMF Working Paper, WP/15/68, pp. 1–26, Available from https://www.imf.org/en/Publications/WP/Issues/2016/12/31/Growing-Un-equal-Fiscal-Policy-and-Income-Inequality-in-China-and-BRIC-42806

Chen, Q., Kang, C., Xia, Q., Zhou, Y., & Yokoyama, R. (2009). Key low-carbon factors in the evolution of power decarbonisation and their impacts on generation expansion planning. *Automation of Electric Power Systems, 15*.

Chen, Y., Liu, S., Wu, H., Zhang, X., & Zhou, Q. (2020). How can Belt and Road countries contribute to glocal low-carbon development? *Journal of Cleaner Production, 256*, 120717.

Cheshmehzangi, A. (2020a). *The city in need: Urban resilience and city management in disruptive disease outbreak events*. Springer.

Cheshmehzangi, A. (2020b). Low carbon transition at the township level: feasibility study of environmental pollutants and sustainable energy planning. *International Journal of Sustainable Energy*, 1–27. https://doi.org/10.1080/14786451.2020.1860042

Cheshmehzangi, A. (2021). From transitions to transformation: a brief review of the potential Impacts of COVID-19 on boosting digitization, digitalization, and systems thinking in the built environment. *Journal of Building Construction and Planning Research, 9*(1), 26–39.

Cheshmehzangi, A., Xie, L., & Tan-Mullins, M. (2018). The role of international actors in low-carbon transitions of Shenzhen's International Low Carbon City in China. *Cities, 74*, 64–74.

Cheshmehzangi, A., Dawodu, A., & Sharifi, A. (2021). *Sustainable urbanism in China*. Routledge.

Climate Action Tracker. (2020). *China going carbon neutral before 2060 would lower warming projections by around 0.2 to 0.3 °C*. Available from https://climateactiontracker.org/press/china-carbon-neutral-before-2060-would-lower-warming-projections-by-around-2-to-3-tenths-of-a-degree/

Dai, S., Zhou, X., & Kuosmanen, T. (2020). Forward-looking assessment of the GHG abatement cost: Application to China. *Energy Economics, 88*, 104758. https://doi.org/10.1016/j.eneco.2020.104758

Dong, L., Gu, F., Fujita, T., Hayashi, Y., & Gao, J. (2014). Uncovering opportunity of low-carbon city promotion with industrial system innovation: Case study on industrial symbiosis projects in China. *Energy Policy, 65*, 388–397.

Edsand, H. E. (2019). Technological innovation system and the wider context: A framework for developing countries. *Technology in Society, 58*, 101150.

EF (Energy Foundation). (2020). *China's new growth pathway: from the 14th five year plan to carbon neutrality* <中国现代化的新征程"十四五"到碳中和的新增长故事>. Available from https://www.efchina.org/Attachments/Report/report-lceg-20201210/Full-Report_Synthesis-Report-2020-on-Chinas-Carbon-Neutrality_EN.pdf

ETC (Energy Transitions Commission). (2019). *China 2050: a fully developed rich zero carbon economy*. Available from http://www.cn-bea.com/filedownload/283476

ETC (Energy Transitions Commission). (2020). *Making mission possible: delivering a net-zero economy*. Available from https://www.teriin.org/sites/default/files/2020-09/Making-Mission-Possible-Report.pdf

ETC (Energy Transitions Commission). (2021). *China zero carbon electricity growth in the 2020s: A vital step toward carbon neutrality*. Available from https://rmi.org/insight/zero-carbon-electricity-growth-in-the-2020s/

Gao, Y. (2016). China's response to climate change issues after Paris climate change conference. *Advances in Climate Change Research, 7*(4), 235–240.

García-García, P., Carpintero, Ó., & Buendía, L. (2020). Just energy transitions to low carbon economies: A review of the concept and its effects on labour and income. *Energy Research & Social Science, 70*, 101664.

Goldman Sachs Research. (2021). *Carbonomics: China net zero—The clean tech revolution*. Available from https://wwwqa.goldmansachs.com/insights/pages/gs-research/carbonomics-china-netzero/report.pdf

Hardeman, W., Sutton, S., Griffin, S., Johnston, M., White, A., Wareham, N. J., & Kinmonth, A. L. (2005). A causal modelling approach to the development of theory-based behaviour change programmes for trial evaluation. *Health Education Research, 20*(6), 676–687.

He, Y., Xu, Y., Pang, Y., Tian, H., & Wu, R. (2016). A regulatory policy to promote renewable energy consumption in China: Review and future evolutionary path. *Renewable Energy, 89*, 695–705.

Hendrickson, T. P., Nikolic, M., & Rakas, J. (2016). Selecting climate change mitigation strategies in urban areas through life cycle perspectives. *Journal of Cleaner Production, 135*, 1129–1137.

Huang, C., Zhang, P., Wang, WL., Huang, Z., Lv, JF., Liu, JZ., & Ni, WD. (2021). The upgradation of coal-fired power generation industry supports China's energy conservation, emission reduction and carbon neutrality. *Thermal Power Generation.*

ICCSD (Institute of Climate Change and Sustainable Development, Tsinghua University). (2020). China low-carbon development strategy and transformation pathways.

Janzen, R., Davis, M., & Kumar, A. (2020). Greenhouse gas emission abatement potential and associated costs of integrating renewable and low carbon energy technologies into the Canadian oil sands. *Journal of Cleaner Production, 272*, 122820. https://doi.org/10.1016/j.jclepro.2020.122820

Jo, J. H., Golden, J. S., & Shin, S. W. (2009). Incorporating built environment factors into climate change mitigation strategies for Seoul, South Korea: A sustainable urban systems framework. *Habitat International, 33*(3), 267–275.

Ju, B-G., Kim, M., Kim, S., Moreno-Ternero, J. D. (2021). Fair international protocols for the abatement of GHG emissions. *Energy Economics, 94*, 105091. https://doi.org/10.1016/j.eneco.2020.105091

Kannan, R. (2009). Uncertainties in key low carbon power generation technologies—Implication for UK decarbonisation targets. *Applied Energy, 86*(10), 1873–1886.

Kennedy, S., & Sgouridis, S. (2011). Rigorous classification and carbon accounting principles for low and zero carbon cities. *Energy Policy, 39*(9), 5259–5268.

Kern, F., & Rogge, K. S. (2016). The pace of governed energy transitions: Agency, international dynamics and the global Paris agreement accelerating decarbonisation processes? *Energy Research & Social Science, 22*, 13–17.

Khanna, N., Fridley, D., Zhou, N., Karali, N., Zhang, J., & Feng, W. (2019). Energy and CO_2 implications of decarbonization strategies for China beyond efficiency: Modeling 2050 maximum renewable resources and accelerated electrification impacts. *Applied Energy, 242*, 12–26.

Lin, W. B., & Zhu, T. (2021). The realization of carbon peak and carbon neutralization need to pay attention to three "Overall Planning". *Price: Theory & Practice.*

Liu, M., & Lo, K. (2020). Pathways to international cooperation on climate governance in China: a comparative analysis. *Journal of Chinese Governance*, 1–18.

Liu, M. P. (2021). Significance, foundation, challenge and policy focus of realizing the goal of "Carbon Neutralization" in China. *Price: Theory & Practice.*

Liu, N. (2018). Will China build a green Belt and Road in the Arctic? *Review of European, Comparative & International Environmental Law, 27*(1), 55–62.

Ma, N., Li, H., Tang, R., Dong, D., Shi, J., & Wang, Z. (2019). Structural analysis of indirect carbon emissions embodied in intermediate input between Chinese sectors: a complex network approach. *Environmental Science and Pollution Research, 26*(17), 17591–17607.

Matsuo, Y., Yanagisawa, A., & Yamashita, Y. (2013). A global energy outlook to 2035 with strategic considerations for Asia and Middle East energy supply and demand interdependencies. *Energy Strategy Reviews, 2*(1), 79–91.

Ministry of Ecology and Environment (MEE), Ministry of Foreign Affairs (MF), National Development and Reform Commission (NDRC) & Ministry of Commerce of PRC. (2017). *Guidance on promoting Green Belt and Road*. <关于推进绿色“一带一路”建设的指导意见>. Available from http://www.mee.gov.cn/gkml/hbb/bwj/201705/t20170505_413602.htm

Ministry of Ecology and Environment (MEE) of PRC. (2017). *Belt and Road ecological and environmental cooperation plan* <“一带一路”生态环保合作规划>. Available from http://www.mee.gov.cn/gkml/hbb/bwj/201705/W020170516330272025970.pdf

Ministry of Commerce. (2013). *Guidelines for environmental protection of foreign investment cooperation* <对外投资合作环境保护指南 >Available from: http://fec.mofcom.gov.cn/article/ywzn/dwtz/zcfg/201512/20151201202217.shtml

Ministry of Commerce. (2015). *Vision and actions on jointly building silk road economic Belt and 21st-century maritime silk road*. <推动共建丝绸之路经济带和21世纪海上丝绸之路的愿景与行动>Available from http://zhs.mofcom.gov.cn/article/xxfb/201503/20150300926644.shtml

Obergassel, W., Lah, O., & Rudolph, F. (2021). Driving towards transformation? To what extent does global climate governance promote decarbonisation of land transport? *Earth System Governance*. https://doi.org/10.1016/j.esg.2021.100098

Paardekooper, S., Lund, H., Chang, M., Nielsen, S., Moreno, D., & Thellufsen, J. Z. (2020). Heat roadmap Chile: A national district heating plan for air pollution decontamination and decarbonisation, *Journal of Cleaner Production, 272*, 122744. https://doi.org/10.1016/j.jclepro.2020.122744

Pan, X., Wang, H., Wang, L., & Chen, W. (2018). Decarbonization of China's transportation sector: in light of national mitigation toward the Paris Agreement goals. *Energy, 155*, 853–864.

Pasimeni, M. R., Valente, D., Zurlini, G., & Petrosillo, I. (2019). The interplay between urban mitigation and adaptation strategies to face climate change in two European countries. *Environmental Science & Policy, 95*, 20–27.

PRI (Principles for Responsible Investment). (2021). *Delivering carbon neutrality in China*. Available from https://www.unpri.org/pri-blog/delivering-carbon-neutrality-in-china/7000.article

Rauland, V., & Newman, P. (2015). *Decarbonising cities: Mainstreaming low carbon urban development*. Springer.

Rio, A. M., & Firza, H. B. (2021). BRI as international norm construction of China's green development strategy in Asia. *Indonesian Journal of International Relations, 5*(1), 1–14.

Sansom, R. (2014). *Decarbonising low grade heat for low carbon future* (Doctoral dissertation). London: Imperial College London.

Shen, C., Jin, J., & Zou, H. F. (2012). Fiscal decentralization in China: history, impact, challenges and next steps. *Annals of Economics & Finance, 13*(1). Available from: http://citeseerx.ist.psu.edu/viewdoc/download?doi=10.1.1.295.9452&rep=rep1&type=pdf

Sorman, A. H., García-Muros, X., Pizarro-Irizar, C., & González-Eguino, M. (2020). Lost (and found) in transition: Expert stakeholder insights on low-carbon energy transitions in Spain. *Energy Research & Social Science, 64*, 101414.

Sovacool, B. K., Baker, L., Martiskainen, M., & Hook, A. (2019). Processes of elite power and low-carbon pathways: Experimentation, financialisation, and dispossession. *Global Environmental Change, 59*, 101985.

Su, B., Heshmati, A., Geng, Y., & Yu, X. (2013). A review of the circular economy in China: moving from rhetoric to implementation. *Journal of Cleaner Production, 42*, 215–227.

Teng, F., & Jotzo, F. (2014). Reaping the economic benefits of decarbonization for China. *China & World Economy, 22*(5), 37–54.

Teo, H. C., Lechner, A. M., Walton, G. W., Chan, F. K. S., et al. (2019). Environmental impacts of infrastructure development under the belt and road initiative. *Environments, 6*(6), 72.

The Oxford Institute for Energy Studies. (2020). *Unpacking China's 2060 carbon neutrality pledge.* Available from https://www.oxfordenergy.org/publications/unpacking-chinas-2060-carbon-neutrality-pledge/

Tyfield, D., Jin, J., & Rooker, T. (2010). *Game-changing China: Lessons from China about disruptive low carbon innovation.* Available from https://eprints.lancs.ac.uk/id/eprint/40880/

Verbong, G. P., Beemsterboer, S., & Sengers, F. (2013). Smart grids or smart users? Involving users in developing a low carbon electricity economy. *Energy Policy, 52*, 117–125.

Wang, C., & Zhang, Y. X. (2020). Implementation pathway and policy system of carbon neutrality vision. *Chinese Journal of Environmental Management,* (06), 58–64. Available from http://zghjgl.ijournal.cn/ch/reader/create_pdf.aspx?file_no=20200611&flag=1&journal_id=zghjgl&year_id=2020

Wang, H., Chen, W., Zhang, H., & Li, N. (2019). Modeling of power sector decarbonization in China: comparisons of early and delayed mitigation towards 2-degree target. *Climatic Change*, 1–14.

WRI (World Resources Institute). (2020). *Accelerating the net-zero transition: strategic action for China's 14th five-year plan.* .Available from https://files.wri.org/s3fs-public/accelerating-net-zero-transition-china_3.pdf?RLUd6F6efcSMSwRt_wI6FT5DcJnYwUfX

XinhuaNet. (2019). *The International Coalition for Green Development on the Belt and Road was established in Beijing to set up a communication platform for green development cooperation.* Available from http://www.xinhuanet.com/world/2019-04/25/c_1124416934.htm

Yam, R. C., Lo, W., Tang, E. P., & Lau, A. K. (2011). Analysis of sources of innovation, technological innovation capabilities, and performance: An empirical study of Hong Kong manufacturing industries. *Research Policy, 40*(3), 391–402.

Yin, W. (2019). Integrating sustainable development goals into the Belt and Road initiative: Would it be a new model for green and sustainable investment? *Sustainability, 11*(24), 6991.

Yong, R. (2007). The circular economy in China. *Journal of Material Cycles and Waste Management, 9*(2), 121–129.

Yu, B. Y., Zhao, G. P., An, R. Y., Chen, J. M., Tan, J. X., & Li, X. Y. (2021). Research on China's CO_2 emission pathway under carbon neutral target. *Journal of Beijing Institute of Technology (Social Sciences Edition).*

Zhang, B., & Wang, Z. (2014). Inter-firm collaborations on carbon emission reduction within industrial chains in China: Practices, drivers and effects on firms' performances. *Energy Economics, 42*, 115–131.

Zhang, H., Geng, Z., Yin, R., & Zhang, W. (2020). Regional differences and convergence tendency of green development competitiveness in China. *Journal of Cleaner Production, 254*, 119922.

Zhang, N., Lior, N., & Jin, H. (2011). The energy situation and its sustainable development strategy in China. *Energy, 36*(6), 3639–3649.

Zhou, L., Gilbert, S., Wang, Y., Cabré, M. M., & Gallagher, K. P. (2018a). *Moving the Green Belt and Road Initiative: From words to actions.* Available from https://www.wri.org/publication/moving-green-belt-and-road-initiative-from-words-to-actions

Zhou, S., Wang, Y., Yuan, Z., & Ou, X. (2018b). Peak energy consumption and CO_2 emissions in China's industrial sector. *Energy Strategy Reviews, 20*, 113–123.

Zou, CN., He, DB., Jia, CY., Xiong, B., Zhao, Q., & Pan, SQ. (2021). Connotation and pathway of world energy transition and its significance for carbon neutral. *Acta Petrolei Sinica.*

Zuo, J., Read, B., Pullen, S., & Shi, Q. (2012). Achieving carbon neutrality in commercial building developments—Perceptions of the construction industry. *Habitat International, 36*, 278–286.

Index

A. Cheshmehzangi and H. Chen, *China's Sustainability Transitions*,
https://doi.org/10.1007/978-981-16-2621-0